3D Printed Smart Sensors and Energy Harvesting Devices

Concepts, fabrication and applications

Online at: https://doi.org/10.1088/978-0-7503-5351-9

3D Printed Smart Sensors and Energy Harvesting Devices

Concepts, fabrication and applications

Edited by

Sanket Goel

Sohan Dudala

MEMS, Microfluidics and Nanoelectronics (MMNE) Lab, Birla Institute of Technology and Science (BITS) Pilani—Hyderabad Campus, Hyderabad, Telangana, India

IOP Publishing, Bristol, UK

ISBN   978-0-7503-5351-9 (ebook)
ISBN   978-0-7503-5349-6 (print)
ISBN   978-0-7503-5352-6 (myPrint)
ISBN   978-0-7503-5350-2 (mobi)

DOI   10.1088/978-0-7503-5351-9

Version: 20241201

IOP ebooks

British Library Cataloguing-in-Publication Data: A catalogue record for this book is available from the British Library.

Published by IOP Publishing, wholly owned by The Institute of Physics, London

IOP Publishing, No.2 The Distillery, Glassfields, Avon Street, Bristol, BS2 0GR, UK

US Office: IOP Publishing, Inc., 190 North Independence Mall West, Suite 601, Philadelphia, PA 19106, USA

The cover image shows a 3D printed electrochemiluminescence (ECL) sensing device intended for biochemical sensing and the corresponding emitted ECL signal. The device was developed using an independent dual extruder FDM 3D printer with PLA and electrically conductive composite filaments in MEMS, Microfluidics and Nanoelectronics (MMNE) Lab, Birla Institute of Technology and Science (BITS) Pilani—Hyderabad Campus, Hyderabad, Telangana, India. The cover image is a derivative of the research work by Abhishek Kumar, a PhD Scholar with MMNE Lab.

Contents

9 Cyber–physical system enabled 3D printed devices

*P Subbulakshmi, L K Pavithra, E Manikandan, K A Karthigeyan
and Inbarani*

10 Applications: smart sensors

*Aniket Chakraborthy, Suresh Nuthalapati, H Harija, Anindya Nag
and Mehment Ercan Altinsoy*

Preface

The advent of 3D printing (3DP) technology, or additive manufacturing (AM), has revolutionised the fabrication and prototyping industry. Recently, 3D printing has become a go-to option for developing next-generation systems and devices including sensors and energy harvesters. This has been possible owing to the development of 3DP technology in terms of new materials and improved performance with high accuracy, robustness and affordability. The devices and systems developed using 3D printing are now integral to emerging technologies, such as micro-electro-mechanical systems (MEMS), nanoelectronics, wearable devices, and microfluidics, which are critical for applications spanning healthcare, energy, and consumer electronics.

This book, *3D Printed Smart Sensors and Energy Harvesting Devices: Concepts, fabrication, and applications*, explores key elements of 3D printing and their transformative potential in sensors and energy harvesters. The book provides a comprehensive look at the entire value chain (cradle to grave) for developing these advanced devices, from design to fabrication, characterisation, and integration with sensing mechanisms and smart circuitry.

Various chapters delve into the technical aspects of 3DP, detailing its advantages over conventional techniques. The readers will gain insight into various 3DP technologies, computer-aided design (CAD) tools, and the importance of materials (including commercially available options and innovative composites). The book discusses how 3DP facilitates diverse applications, from sensing, energy storage and harvesting to fuel cell and supercapacitor encapsulation, by integrating functional materials such as electrodes and employing chemical modifications.

This book offers a forward-looking perspective, discussing challenges and future opportunities for 3DP devices. The book also highlights the interdisciplinary nature of realising devices and systems and the role of 3DP technology in addressing their integration challenges. 3DP and associated technologies have bridged the gap between engineering, material science, and applied research, making it a valuable tool for students, academics, industry professionals, or anyone interested in the forefront of manufacturing innovation and prototyping.

We invite readers to explore the concepts and techniques for developing smart sensors and energy-harvesting devices employing 3D printing. The book is designed to keep in view the novice needs of readers, with initial chapters discussing the fundamentals and subject matter experts who may be interested in reading about recent advancements.

Acknowledgements

This book is a result of contributions and support from many individuals. I would like to express my sincere gratitude to everyone who has helped to bring this project to completion.

First and foremost, I would like to thank the authors of each chapter for their willingness to share their knowledge and expertise. Their contributions have made this book a comprehensive and valuable resource for anyone interested in 3D printing technologies for sensors, energy harvesters and allied systems.

I would also like to express my gratitude to Dr John Navas, Phoebe Hooper and the entire production team from IOP Publishing for their invaluable assistance with the publication process. Their dedication and hard work have been instrumental in completing this book.

Finally, I would like to thank the members of our MEMS, Microfluidics & Nanoelectronics (MMNE) Lab, who have time and again done inspirational work using 3D printing technologies, which led to the conceptualisation of this book.

Sanket Goel, PhD

Editor biographies

Sanket Goel

Sanket Goel is a Professor with the Department of Electrical and Electronics Engineering at BITS Pilani, Hyderabad Campus. He previously headed the same department and R&D department at UPES, Dehradun. He is the Principal Investigator of MEMS, Microfluidics and Nanoelectronics (MMNE) Lab, which works towards realising futuristic smart sensors and intelligent energy harvesters encompassing various multidisciplinary domains. Professor Goel has published >350 scientific articles in various domains, including microfluidics, biosensors, nanoelectronics, fuel cells, smart sensors, MEMS, solar energy, wearable devices and cyber-physical systems. He has 60+ patents (including 12 granted) to his credits, has delivered >110 invited talks and has supervised 50+ PhD students. Professor Goel is also on the editorial board of numerous journals and has a multitude of accolades including Fulbright and JSPS fellowships. Currently, he is also the Dean (Research and Innovation) at BITS Pilani and Distinguished Lecturer of IEEE Sensors Council. Professor Goel is a co-founder of three spin-offs: Cleome Innovations, Pyrome Innovations, and Sensome Innovations.

Sohan Dudala

Sohan Dudala completed his doctoral degree from the Department of Electrical and Electronics Engineering, BITS Pilani, Hyderabad Campus, in the field of bio-MEMS and microfluidics. Prior to his doctoral program, he completed an extra credit program with Hiroshima University. Sohan also holds a master's degree in mechanical engineering from BITS Pilani, Hyderabad Campus and a bachelor's degree in mechanical engineering from Manipal University Jaipur. Sohan bears wide exposure to interdisciplinary technology development with his previous work in the fields of metal oxides nano-structure coatings; microfluidic water and soil assessment systems; microfluidic cell culture; PCR diagnostic devices; and point-of-care diagnostics. His current research interests include bio-microfluidics, bio-MEMS and wireless communication.

List of contributors

Shanta Kumari Adiki
Rajarshi Shahu College of Pharmacy, Markhel, Nanded 431718, Maharashtra, India

Mehment Ercan Altinsoy
Faculty of Electrical and Computer Engineering, Technische Universität Dresden, Dresden 01062, Germany
and
Centre for Tactile Internet with Human-in-the-Loop (CeTI), Technische Universität Dresden, Dresden 01069, Germany

Ravi Kumar Arya
Xiangshan Laboratory, Zhongshan Institute of Changchun University of Science and Technology, Zhongshan, Guangdong, China

Suresh Balpande
Department of Information Technology and Security, Ramdeobaba University, Nagpur 440013, Maharashtra, India

Ayan Bhatnagar
Department of Mechanical Engineering, Birla Institute of Technology and Science, Pilani—Pilani Campus, India

Madhavi Lakshmi Ratna Bhavaraju
Dhanwantari College of Pharmacy, Chikkabanavara Bengaluru 560090, India

Aniket Chakraborthy
Faculty of Electrical and Computer Engineering, Technische Universität Dresden, Dresden 01062, Germany
and
Centre for Tactile Internet with Human-in-the-Loop (CeTI), Technische Universität Dresden, Dresden 01069, Germany

Apurba Das
Department of Aerospace Engineering, IIEST Shibpur, Howrah 711103, India

Junwei Dong
Xiangshan Laboratory, Zhongshan Institute of Changchun University of Science and Technology, Zhongshan, Guangdong, China

T Lachana Dora
Department of Mechanical Engineering, Birla Institute of Technology and Science, Pilani—Pilani Campus, India

Sohan Dudala
MEMS, Microfluidics and Nanoelectronics (MMNE) Lab, Birla Institute of Technology and Science (BITS) Pilani—Hyderabad Campus, Hyderabad, Telangana, India

Sanket Goel
MEMS, Microfluidics and Nanoelectronics (MMNE) Lab, Birla Institute of Technology and Science (BITS) Pilani—Hyderabad Campus, Hyderabad, Telangana, India

Sarang P Gumfekar
Department of Chemical Engineering, Indian Institute of Technology Ropar, Rupnagar, Punjab 140001, India

H Harija
Faculty of Electrical and Computer Engineering, Technische Universität Dresden, Dresden 01062, Germany

Inbarani
Centre for Innovation and Product Development, Vellore Institute of Technology, Chennai, India
and
School of Electronics Engineering, Vellore Institute of Technology, Chennai, India

Ashish Kapoor
Department of Chemical Engineering, Harcourt Butler Technical University, Kanpur, Uttar Pradesh 208002, India

K A Karthigeyan
Department of Electronics and Communication Engineering, Vel Tech Rangarajan Dr Sagunthala R&D Institute of Science and Technology, Chennai 600062, India

Prakash Katakam
Indira College of Pharmacy, Nanded 431606, Maharashtra, India

M Junaid Khan
Modern Arch Infrastructure Pvt Ltd, Nagpur, India

R S Krishna
Department of Mechanical Engineering, Birla Institute of Technology and Science, Pilani—Hyderabad Campus, India
and
Centre for Infrastructure Engineering, Western Sydney University, Penrith, NSW 2751, Sydney, Australia

Abhishek Kumar
MEMS, Microfluidics and Nanoelectronics (MMNE) Lab, Birla Institute of Technology and Science (BITS) Pilani—Hyderabad Campus, Hyderabad, Telangana, India

Pawan Kumar
Calibyte Private Limited, India

Rakesh Kumar
Department of Mechanical Engineering, Indian Institute of Technology Ropar, Rupnagar, Punjab 140001, India

E Manikandan
Centre for Innovation and Product Development, Vellore Institute of Technology, Chennai, India
and
School of Electronics Engineering, Vellore Institute of Technology, Chennai, India

Valentin Mateev
Department of Electrical Apparatus, Technical University of Sofia, Bulgaria

Radhan Raman Mishra
Department of Mechanical Engineering, Birla Institute of Technology and Science, Pilani—Pilani Campus, India

Arnab Mukherjee
Department of Mechanical Engineering, Netaji Subhash Engineering College, Kolkata 700152, India

Anindya Nag
Faculty of Electrical and Computer Engineering, Technische Universität Dresden, Dresden 01062, Germany
and
Centre for Tactile Internet with Human-in-the-Loop (CeTI), Technische Universität Dresden, Dresden 01069, Germany

Gajanan Nikhade
Department of Mechanical Engineering, Ramdeobaba University, Nagpur 440013, Maharashtra, India

Suresh Nuthalapati
Faculty of Electrical and Computer Engineering, Technische Universität Dresden, Dresden 01062, Germany
and
6G-life Research Hub, Technische Universität Dresden, Dresden 01062, Germany

L K Pavithra
School of Computer Science and Engineering, Vellore Institute of Technology, Chennai, India

Amrita Priyardishini
Department of Mechanical Engineering, Birla Institute of Technology and Science, Pilani—Hyderabad Campus, India

Srinivasa Prakash Regalla
Department of Mechanical Engineering, Birla Institute of Technology and Science-Pilani, Hyderabad Campus, India

Suman Saha
Department of Civil Engineering, National Institute of Technology Durgapur, West Bengal, India

Venkateswaran Pedinti Sankaran
Department of Academics and Special Projects, CADFEM India Private Limited, Chennai, India

Arijit Sinha
Department of Metallurgical Engineering, Kazi Nazrul University, Asansol 713 340, India

P Subbulakshmi
School of Computer Science and Engineering, Vellore Institute of Technology, Chennai, India

Kurra Suresh
Department of Mechanical Engineering, Birla Institute of Technology and Science, Pilani—Hyderabad Campus, India

Aswin Chowdary Undavalli
Electrical and Systems Engineering Washington University in St Louis, Missouri, 63130, USA

Vijay Vaishampayan
Department of Chemical Engineering, Indian Institute of Technology Ropar, Rupnagar, Punjab 140001, India

Shrikant Vidya
School of Mechanical Engineering, Galgotias University, Greater Noida 201310, India

Adil Wazeer
School of Laser Science and Engineering, Jadavpur University, Kolkata 700032, India

IOP Publishing

3D Printed Smart Sensors and Energy Harvesting Devices
Concepts, fabrication and applications
Sanket Goel and Sohan Dudala

Chapter 1

Introduction to 3D printing

Abhishek Kumar and Sanket Goel

Over the past two decades, 3D printing, or additive manufacturing, has drawn a lot of attention from both academia and industry. It has rapidly evolved from the prototype development process to the product development process. The challenges faced by traditional production are being overcome by the ongoing progress in materials for the additive manufacturing process. It is emerging as the reliable solution for fabricating intricate and customized complex tailored geometries unreachable with subtractive manufacturing. Its key advantages include cutting initial costs, adopting innovative design methods, and aligning with cutting edge technologies. The presented chapter introduces the background of the additive manufacturing process, its principle, its layer-by-layer part-building mechanism, its classification, and its advancements. The subsequent chapters will provide detailed insights encompassing fundamental aspects, advanced principles, and comprehensive analysis, thus providing a thorough understanding of the topic.

1.1 Introduction

The fabrication of objects using 3D printing has been a subject of considerable intrigue among technology entrepreneurs, science fiction writers, the media, and economic forecasts [1–4]. Whether miniaturised devices, prosthetic arms, or sophisticated wearable electronic devices, 3D printing is employed in several diversified applications. Additive manufacturing, or industrial 3D printing, is a sequential layer-by-layer 3D fabrication process that directly uses computer aided design (CAD) model data without requiring direct tooling or jig and fixtures to create complex geometries [5–7]. Additive manufacturing differs from subtractive (lathe, milling and turning) and formative (casting or forging) manufacturing processes in several aspects such as material removal for the creating part, limited ability to create complex structures, wide availability of materials, and inability to achieve zero lead time due to requirement of tooling (jigs and fixtures). Additive manufacturing is rapidly advancing towards sustainability and allowing the user to choose

doi:10.1088/978-0-7503-5351-9ch1
1-1

various materials based on their desired applications, such as metal, polymer, ceramics, bio-gels, and novel composites [8–10]. Additive manufacturing processes offer a multitude of benefits, including but not limited to minimal material waste, decreased expense related to machine setup, the capability to fabricate 3D geometries featuring intricate details and architecture, and customization.

In the 1980s, when the first additive manufacturing technology was introduced to the market, it was known by several names, including layer-by-layer printing, generative manufacturing, rapid prototyping, freeform manufacturing, and so on. Oftentimes, it was referred to as 3D printing. Despite its ongoing development, this technology has demonstrated enormous potential to contribute significantly to intelligent connected digital industrial production [4, 11–16].

The digitalization of the manufacturing process has altered the way in which industries operate. The impetus behind every industrial revolution has always been a manufacturing process vital for socioeconomic development. Industrial revolution 1.0 started at the end of the eighteenth century and mobilized industries using power, steam, and water to mechanize production. Industrial revolution 1.0 is considered a significant transition phase for sectors today and the most remarkable breakthrough for improving industrial and human productivity. Industrial revolution 2.0 facilitated the development of electric power, assembly line systems, and mass production in the nineteenth century. Beginning in the twentieth century, industrial revolution 3.0 automated a portion of the manufacturing processes with the aid of simple computers and programmable memory controls. With the development of intelligent factories, digital twins, and cutting-edge information systems, we are ushering in the fourth industrial revolution in our contemporary world. Industrial revolution 4.0 offered numerous benefits, dramatically transiting industries digitally by pulling people into more innovative network-machine connectivity. Additive manufacturing is a crucial enabler of industrial revolution 4.0 because it allows complete customization and provides enterprises with the ability to fulfil product design demands as per customer needs and individual requirements [1, 2, 17–19].

1.2 Generic part-building workflow of the additive manufacturing process

From design ideation to product realization, all established additive manufacturing processes follow the part-building workflow.

1.2.1 Stage 1. Design conceptualization and design preparation

The key to successful product fabrication is design creativity and planning to know how the product will function, its technical consideration, economic consideration, and aesthetics. Preparing error-free 3D model data is the primary prerequisite for additive manufacturing workflow. The product design phase can start from textual narrative description to representative 3D models to realize the final prototype or product. Alternatively, reverse engineering may be used by 3D scanners and image data acquisition systems to prepare the 3D CAD model by deriving relevant measurements from point clouds from the surface. The representative 3D model may be created using professional CAD software to prepare relationally structured 3D CAD solid models.

1.2.2 Stage 2. Conversion of the 3D CAD model to an .STL file and .STL file manipulation

Standard Triangulation Language, or Surface Tessellation Language (.STL), has been established as the standard file extension for any generic additive manufacturing process workflow. The prepared 3D solid CAD model is converted to .STL file format to remove design history (modelling data and geometry construction information) with approximate series of triangular mesh facets. STL files do not represent any measurement unit, geometry feature, materials, geometry function, or design colour; it is a collection of unordered triangular mesh showing normal surface vectors. STL is a straightforward file format since it only represents the surface vector description of the prepared 3D CAD model—the primary advantage of using the .STL file format in any additive manufacturing process workflow allows users to convert a ready 3D CAD model into layer slices for the additive manufacturing process according to defined slicing parameters. Further, the STL file is checked for manifolds in geometries to ensure watertight geometry and optimized for incorporating the slicing parameters. Various parameters, such as layer height, raster angle, width, infill pattern, infill density air gap, shell thickness, material printing properties, machine set-up information, part orientation, etc, are essential for preparing code per additive manufacturing machines.

1.2.3 Stage 3. Additive manufacturing part preparation and post-processing

In this stage, additive manufacturing machine technology is set up based on the specified parameters, material loading, essential components, build plate, or chamber calibration respective to machine axes. After carrying out necessary machine functions, the part-building may be started by a computer-controlled sequential layer deposition mechanism to build the final product. Before the part-building process, the mechanical engineer must load the build material in the specified chamber. The final build product may undergo clean-up and post-processing. The post-processing of the building part may involve chemical and thermal treatment to get accurate, precise, mechanical stress-free, ready-to-use geometry. The post-processing step may vary based on the utilized additive manufacturing technology.

1.3 Materials for additive manufacturing

The additive manufacturing process has revolutionized conventional manufacturing by changing the fabrication principle and product development phase [8, 9]. The advancement and ongoing research for high-strength composites for the additive manufacturing process have overcome the restriction of the additive manufacturing process and have been a breakthrough in the advanced manufacturing process. The wide variety of materials provides a strong foundation for additive manufacturing.

Currently, the additive manufacturing process uses metal, metal-based composites, polymer, and polymer-based composites, ceramics, resins, and bio-gels for product fabrication [10, 20].

1.3.1 Ceramic

Ceramic materials are generally defined as compounds that contain metal oxides, borides, nitrides, and carbides. Several ceramic materials, such as clay, aluminium oxide, bioceramics, and titanium oxide, are available for additive manufacturing. 3D printing of ceramic materials remains challenging, as monitoring the fabrication process precisely and controlling the shrinkage during fabrication is necessary. The ceramics-based 3D printed part has been employed for various structural and non-structural applications due to their unique features of characteristics such as mechanical, biocompatibility, thermal, and chemical properties [21, 22].

1.3.2 Polymer

Polymeric materials are synthetic compounds that could change shape during processing. There are a wide variety of polymeric materials (thermoplastics and thermosetting) available for 3D printing, such as polyethene terephthalate (PET), acrylonitrile butadiene styrene (ABS), poly-lactic acid, polyethene (P.E.), and poly-propylene (P.P.) and their composites. Using polymers in 3D printing provides an opportunity to recycle biodegradable plastics. Recycled polymeric waste has been the subject of extensive research recently, with the goal of increasing sustainability through the fabrication of both functional and non-functional prototypes [10, 23–25].

1.3.3 Metals

The use of metal in 3D printing has been widely reported for biomedical and electronics applications. Many metals are available for 3D printing, such as stainless steel, titanium, copper, aluminium, tin, copper, nickel, cobalt, etc. Fabricating metallic components with additive manufacturing provides numerous advantages compared to subtractive manufacturing, such as design freedom to print complex structures, reduction in production time, improved structural properties, and topology optimisation [26, 27].

1.4 Classification of additive manufacturing technologies

According to ASTM F2792 standards, additive manufacturing technologies are classified into seven process families (figure 1.1) [28].

1.5 Powder bed fusion

In a powder bed fusion (PBF) system, a laser or an electron beam is used to selectively fuse powdered material on the build plate to fabricate parts [29]. PBF has derivatives such as selective laser melting (SLM), selective laser sintering (SLS), and direct metal laser sintering (DMLS). It provides numerous functionalities to users for fabricating

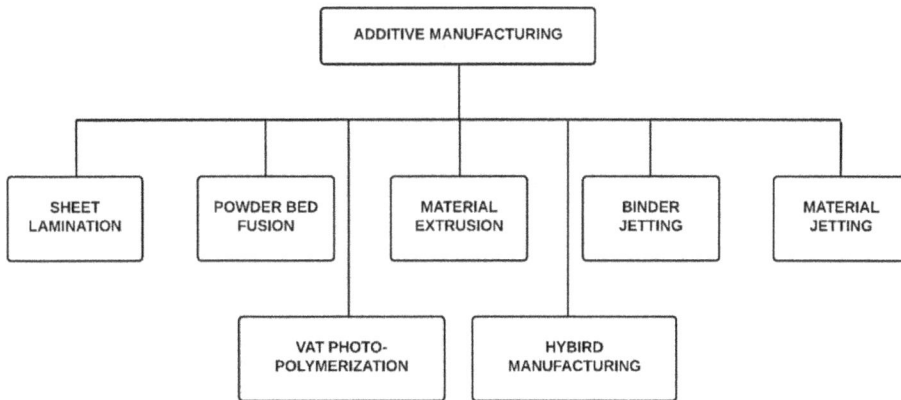

Figure 1.1. Classification of the additive manufacturing process.

bespoke metallic components in widespread engineering applications. The PBF system is mainly comprised of five main essential parts: (i) build platform, (ii) power source, (iii) material reservoir, (iv) mechanical roller and (v) material collector [21, 30]. This process requires secondary support for part fabrication, which also makes it suitable for creating complex geometries. This process is extensively used in medical, manufacturing, and aerospace applications. This fabrication process is a thermally driven process, which needs to be modelled precisely [24, 31, 32].

1.6 Material extrusion

The material extrusion class of additive manufacturing technology belongs to the part fabrication process. It is achieved by forcing feedstock material in a semi-molten state from a heated extruder head and depositing it sequentially controlled layer-by-layer fashion according to CAD model data on the heated build platform [19]. The widely used material extrusion process is fused deposition modelling (FDM), alternatively known as fused filament fabrication. The general phenomenon of the material extrusion process involves the melting of polymer and polymer-based matrix or extruding material slurries through syringe dispensing. Material extrusion-based processes are currently the most popular and widely used technology in the market over other available additive manufacturing technology. The primary approach in the material extrusion process is controlling the material state using temperature inside the liquefier head so that the molten material will flow out through the nozzle head and will get fused with the adjacent deposited material on the heated build platform [25, 33, 34].

1.7 Binder jetting

Binder jetting is an additive manufacturing process used to fabricate parts. Using liquid binding in predefined regions, parts are built layer by layer until the position is complete. Binder jet printing provides numerous advantages due to the use of binder in part fabrication over the other existing additive manufacturing process like creating coloured geometries using coloured binders, reduction of heat affected zone, and capabilities to print a wide range of materials such as metals, ceramics, sand, and polymers. It has been employed for dentistry, foundry, and biomedicine applications [35]. This process lacks quantitative and qualitative characteristics between the printed parts due to attributed uncertainties such as inadequate process monitoring. This process requires controlled monitoring for proper part fabrication. The appropriate part fabrication with binder jetting also requires process optimization for build plate interaction with the binder saturation level (droplet velocity impact, binder size distribution, and mean size of the particle) [27, 36, 37].

1.8 Sheet lamination

In the sheet lamination additive manufacturing process, thin sheets of materials are fed through a cylindrical roller and bonded together layer by layer to form a single piece of a part. There are two types of sheet lamination additive manufacturing processes: (i) laminated object manufacturing, and (ii) ultrasonic consolidation. Here, various materials, such as paper, polymer, metal etc can be used. The chosen material will determine the process characteristics of the sheet lamination process. The major drawback of this fabrication process is that, in order to produce parts of the proper quality, the process must be more accurate and precise [10, 16, 23, 38]. The process starts with positioning the selected material over the build plate and bonding over the other layer with adhesive and pressure. Then the next layer is added, and extra fabric is cut using a knife or laser following the computer-controlled path. It is one of the most straightforward 3D printing part fabrication processes, and has low cost per part fabricated [6, 7, 14].

1.9 VAT photopolymerization

VAT photopolymerization is an additive manufacturing process that uses liquid photopolymer resin to create a geometry layer by layer. The photopolymer resins are hardened using ultraviolet light wherever required. In this process, various parts are fabricated downwards without using structural support from the material during the part-build process [20, 39]. This process is categorised into three types: (i) stereo-lithography, (ii) direct light processing, and (iii) masked stereolithography. The photopolymer resin used in this process is light-activated material, which changes its properties when exposed to light. Regarding quality, it is a high-accuracy part-printing process with details. This process has limitations, such as using an expensive part-printing process, inadequate strength, low durability, etc [10, 25].

1.10 Material jetting

In the material jetting additive manufacturing process, different parts are fabricated with the spread of tiny material droplets across the build plate and cured with ultraviolet (UV) light. The build plate drops substantially downward when the preceding layer is deposited [4, 39, 40]. This process continues until the whole geometry is fabricated. The main critical components of the material jetting machine setup are: (i) print heads for the deposition of material across the build platform, (ii) the build platform, where the material is deposited and moved downwards according to the height of the material, (iii) material source for storing the material and feeding material through the process and (iv) light source. A UV light source is located near the print heads for solidifying the structure. It is also named polyjet printing and nanoparticle jetting. It can provide very high accuracy and low wastage of material due to accurate jetting over the required region. However, part printing with this technology lacks mechanical properties and additional material for the support structure [9, 19, 41].

1.11 Hybrid manufacturing

In hybrid manufacturing, materials are fabricated by the direct energy deposition method with a computer numerical control machining step involving material in wire or powder form and material fusion using a laser as an energy source. The hybrid manufacturing process combines additive and subtractive manufacturing in a single-machine setup. This process sometimes involves using a robotic arm which also makes this process suitable for fabricating complex geometries [7, 42, 43]. This process provides advantages such as a high surface finish with geometrical freedom, in-process support removal, etc. Hybrid manufacturing provides opportunities for fabricating products efficiently and rapidly in a unified environment with fewer errors.

1.12 Design for additive manufacturing for quality

Design for manufacturing and assembly (DfAM) is the most widespread manufacturing practice used by manufacturing and design engineers for tailoring the design to avoid manufacturing difficulties and to minimize costs in conventional manufacturing processes. The functionalities of the additive manufacturing process provide the means to rethink the design for the manufacturing and assembly approach. Integrating the DfAM approach with additive manufacturing requires extensive knowledge of material behaviour, assembly process, machine capabilities, process capabilities, supplier capabilities, etc. The DfAM approach looks conceptually simple but can be complex and time-consuming to integrate with the new manufacturing process. The induction of the DfAM flow in additive manufacturing may reduce the lead time for mass production and delivering complex geometries. Incorporating DfAM with additive manufacturing will maximize product and machine performance in shape complexity, design synthesis, material composition, etc. The well-planned additive manufacturing process has the potential to be a great tool. The core concept of the DfAM process lies with the planned engineering applications and function of the intended product.

Since additive manufacturing is a carefully thought-out process, there is a significant amount of proportionality between the design and the anticipated performance of the product. A well-planned additive manufacturing process provides the ability to fabricate complex lattice structures unattainable by the conventional manufacturing process. Although extensive research has been reported on developing new materials for additive manufacturing, the commercialization and mass production of additively manufactured components will be determined by a well-planned design for the additive manufacturing process. It is understandable that the product development cycle will be expensive but must be balanced with excellent products in design, product quality, product feasibility, and process feasibility.

1.13 Conclusion and discussion

Sustainability is the key to the success of any manufacturing process in achieving economic, technical, and environmental feasibility. These days, industries, researchers, and academics are concentrating on transforming additive manufacturing into a sustainable production method. Design ideation and prototype realization are part of the product development phase, and they are conceptualized after market and user needs are understood. With the adoption of additive manufacturing processes across various industries, there is a significant need to educate the workforce with a good understanding of the additive manufacturing process. University courses and training programmes for additive manufacturing should be developed, according to the roadmap for educating the workforce with advanced processes. Generative design and reverse engineering are iterative processes for developing a new design or redesigning an existing product.

Additive manufacturing will dominate the existing mass manufacturing technique. It has attracted unparalleled attention across the world in widespread applications. The planning and optimization of the additive manufacturing process have been crucial for process stability leading to the generation of relevant part characteristics, shape characteristics, and degree of design freedom with certain fidelity. Concurrent manufacturing process analysis, model development, and simulation approaches are limited in the additive manufacturing process. Integrating real-time processes and feedback loop monitoring may bring a breakthrough in the new product realization process.

The additive manufacturing process has witnessed rapid growth in aerospace, manufacturing, food engineering, biomedical engineering, and electronic industries. It has shifted from a prototype to a product development process, increasing its scalability and overcoming the challenges of the existing conventional manufacturing process. However, the additive manufacturing process still faces challenges: repeatability, part yield rate, ability to deploy in mass production, and material wastage in the support structure. The following chapters will provide detailed insights encompassing fundamental aspects, advanced principles, and comprehensive analysis, thus providing an understanding of the topic thoroughly.

References

[1] Singh R, Kumar A and Boparai K S 2022 Intramedullary pin fixation in 3D printed canine femur bone model for preoperative surgical planning *J. Brazil. Soc. Mech. Sci. Eng.* **44** 299
[2] Singh R, Barwar A and Kumar A 2022 Investigations on primary and secondary recycling of PLA and its composite for biomedical and sensing applications *J. Inst. Eng. (India)* C **103** 821–36

[3] Haleem A and Javaid M 2020 3D printed medical parts with different materials using additive manufacturing *Clin. Epidemiol. Glob. Health* **8** 215–23

[4] Ashima R, Haleem A, Bahl S, Javaid M, Mahla S K and Singh S 2021 Automation and manufacturing of smart materials in additive manufacturing technologies using Internet of Things towards the adoption of industry 4.0 *Mater. Today Proc.* **45** 5081–8

[5] Gibson I, Rosen D, Stucker B and Khorasani M 2021 *Additive Manufacturing Technologies* (Berlin: Springer)

[6] Eyers D R and Potter A T 2017 Industrial additive manufacturing: a manufacturing systems perspective *Comput. Ind.* **92–93** 208–18

[7] Renjith S C, Park K and Okudan Kremer G E 2020 A design framework for additive manufacturing: integration of additive manufacturing capabilities in the early design process *Int. J. Precis. Eng. Manuf.* **21** 329–45

[8] Bourell D *et al* 2017 Materials for additive manufacturing *CIRP Ann.* **66** 659–81

[9] Gibson I, Rosen D, Stucker B and Khorasani M 2021 Materials for additive manufacturing *Additive Manufacturing Technologies* (Berlin: Springer) pp 379–428

[10] Jiaying Tan L, Zhu W, Zhou K, Tan L J, Zhu W and Zhou K 2020 Recent progress on polymer materials for additive manufacturing *Adv. Funct. Mater.* **30** 2003062

[11] Chen L, He Y, Yang Y, Niu S and Ren H 2017 The research status and development trend of additive manufacturing technology *Int. J. Adv. Manuf. Technol.* **89** 3651–660

[12] Ji L and Zhou T 2011 Predicting product precision in fused deposition modelling based on artificial neural network *Adv. Sci. Lett.* **4** 2193–7

[13] Adekanye S A, Mahamood R M, Akinlabi E T and Owolabi M G 2017 Additive manufacturing: the future of manufacturing: Dodajalna (3D) Tehnologija: Prihodnost Proizvjanja *Mater. Tehnol.* **51** 709–15

[14] Abdulhameed O, Al-Ahmari A, Ameen W and Mian S H 2019 Additive manufacturing: challenges, trends, and applications *Adv. Mech. Eng.* **11**

[15] Valilai O F and Houshmand M 2015 Depicting additive manufacturing from a global perspective; using cloud manufacturing paradigm for integration and collaboration *Proc. Inst. Mech. Eng.* B **229** 2216–37

[16] Jin Y, Ji S, Li X and Yu J 2017 A scientometric review of hotspots and emerging trends in additive manufacturing *J. Manuf. Technol. Manage* **28** 28–38

[17] Boparai K S, Kumar A and Singh 2022 Primary and secondary melt processing for plastics *Additive Manufacturing for Plastic Recycling: Efforts in Boosting A Circular Economy* (Boca Raton, FL: CRC Press) pp 51–65

[18] Singh R, Kumar A and Boparai K S 2022 Additive manufacturing assisted preoperative surgical planning for canine femur bone fracture *Natl Acad. Sci. Lett.* **45** 521–4

[19] Boparai K S, Kumar A and Singh R 2022 On characterization of rechargeable, flexible electrochemical energy storage device *4D Printing: Fundamentals and Applications* (Amsterdam: Elsevier) pp 67–88

[20] Ghanem M A *et al* 2020 The role of polymer mechanochemistry in responsive materials and additive manufacturing *Nat. Rev. Mater.* **6** 84–98

[21] Mussatto A *et al* 2022 Laser-powder bed fusion in-process dispersion of reinforcing ceramic nanoparticles onto powder beds via colloid nebulization *Mater. Chem. Phys.* **287** 126245

[22] Zocca A, Colombo P, Gomes C M and Günster J 2015 Additive manufacturing of ceramics: issues, potentialities, and opportunities *J. Am. Ceram. Soc.* **98** 1983–2001

[23] Park S and Fu K 2021 Polymer-based filament feedstock for additive manufacturing *Compos. Sci. Technol.* **213** 108876

[24] Sillani F, Schiegg R, Schmid M, Macdonald E and Wegener K 2022 Powder surface roughness as proxy for bed density in powder bed fusion of polymers *Polymers (Basel)* **14** 81

[25] Jiang K Y and Gu Y H 2004 Controlling parameters for polymer melting and extrusion in FDM *Key Eng. Mater.* **258–259** 667–71

[26] Xia Y, Dong Z-W, Guo X-Y, Tian Q-H and Liu Y 2020 Towards a circular metal additive manufacturing through recycling of materials: a mini-review *J. Cent. South Univ.* **27** 1134–45

[27] Bai Y and Williams C B 2018 The effect of inkjetted nanoparticles on metal part properties in binder jetting additive manufacturing *Nanotechnology* **29** 395706

[28] ASTM International 2012 Standard Terminology for Additive Manufacturing Technologies https://astm.org/f2792-12.html (accessed 9 December 2022)

[29] Ogawahara M and Sasaki S 2021 Effects of defects and inclusions on the fatigue properties of inconel 718 fabricated by laser powder bed fusion followed by HIP *Mater. Trans.* **62** 631–5

[30] Jacob G, Donmez A, Slotwinski J and Moylan S 2016 Measurement of powder bed density in powder bed fusion additive manufacturing processes *Meas. Sci. Technol.* **27** 115601

[31] Kouprianoff D 2021 Investigation of acoustic emission signal during laser powder bed fusion at different operating modes *South Afr. J. Indust. Eng.* **32** 279–83

[32] Amiri M and Payton E J 2021 An analytical model for prediction of denudation zone width in laser powder bed fusion additive manufacturing *Addit. Manuf.* **48** 102461

[33] Butt J, Bhaskar R and Mohaghegh V 2021 Investigating the effects of extrusion temperatures and material extrusion rates on FFF-printed thermoplastics *Int. J. Adv. Manuf. Technol.* **117** 2679–99

[34] Darwish L R, El-Wakad M T and Farag M M 2021 Towards an ultra-affordable three-dimensional bioprinter: a heated inductive-enabled syringe pump extrusion multifunction module for open-source fused deposition modeling three-dimensional printers *J. Manuf. Sci. Eng.* **143** 125001

[35] Ziaee M and Crane N B 2019 Binder jetting: A review of process, materials, and methods *Addit. Manuf.* **28** 781–801

[36] Mirzababaei S, Paul B K and Pasebani S 2022 Microstructure-property relationship in binder jet produced and vacuum sintered 316 L *Addit. Manuf.* **53** 102720

[37] Hartmann C, van den Bosch L, Spiegel J, Rumschöttel D and Günther D 2022 Removal of stair-step effects in binder jetting additive manufacturing using grayscale and dithering-based droplet distribution *Materials* **15** 3798

[38] Bandyopadhyay A and Heer B 2018 Additive manufacturing of multi-material structures *Mater. Sci. Eng.* R **129** 1–16

[39] Oropeza D, Roberts R and Hart A J 2022 A rapid development workflow for binder inks for additive manufacturing with application to polymer and reactive binder ink formulation *J. Manuf. Process* **73** 471–82

[40] Ning F, Cong W, Hu Z and Huang K 2017 Additive manufacturing of thermoplastic matrix composites using fused deposition modeling: A comparison of two reinforcements *J. Compos. Mater.* **51** 3733–42

[41] Miao G, Du W, Pei Z and Ma C 2022 A literature review on powder spreading in additive manufacturing *Addit. Manuf.* **58** 103029

[42] Li L, Haghighi A and Yang Y 2018 A novel 6-axis hybrid additive-subtractive manufacturing process: design and case studies *J. Manuf. Process.* **33** 150–60

[43] Awasthi P and Banerjee S S 2021 Fused deposition modeling of thermoplastic elastomeric materials: challenges and opportunities *Addit. Manuf.* **46** 102177

IOP Publishing

3D Printed Smart Sensors and Energy Harvesting Devices
Concepts, fabrication and applications
Sanket Goel and Sohan Dudala

Chapter 2

Types of 3D printing techniques

Venkateswaran Pedinti Sankaran

With the advent of rapid prototyping combined with the limitless possibilities of 3D printing, the field of smart sensors and microfluidic devices is now in its golden era of operation. Energy harvesting devices require extremely low sample volumes, and have tiny integrated systems and low measurement times, making them ideal diagnostic tools for the biomedical world. Such micro-total-analysis systems can conduct parallel processing of tasks simultaneously to produce the required outputs. The application of rapid manufacturing technologies is useful for various functions ranging from biological studies to various aspects of energy, electronics, chemistry, and life sciences. This chapter covers the use of rapid fabrication techniques based on additive manufacturing. Components of optical and electronic devices, such as valves, mixers, and infusion pumps, can be realized using rapid prototyping techniques. These techniques include vat photopolymerization, stereolithography, digital light processing, fused deposition modelling, material jetting, powder bed fusion, selective laser sintering, and electron beam melting. Because of their short fabrication time and low-cost materials, these methods have a lower cost, and are highly effective and efficient compared to the conventional micromachining used for MEMS, which is, however, very stable and robust with a high reproducibility rate. The drawbacks of micromachining are the time taken for fabrication, the amount of material consumed, and the cost of manufacturing.

2.1 Types of 3D printing techniques

2.1.1 Classification of 3D printing techniques

3D printing, first patented in the year 1986 and also known as additive manufacturing or rapid prototyping, encompasses a wide spectrum of techniques. Material is either deposited, joined, or solidified to create a physical object from a digital file after pre-processing. The advent of this technique has led to a situation in which a vast majority of the available technologies are currently open-source and major players in the market have begun to explore the spectrum to diversify their portfolio to expand their customer base. The following is a generalized classification of various 3D printing processes used globally [1–3]:

- Binder jetting
- Direct energy deposition
- Material extrusion
- Material jetting
- Powder bed fusion
- Sheet lamination
- Vat polymerization

Looking deeper, the realization of all the above technologies can be further narrowed down by commercial names and terms, leaving us with a broader classification chart wherein the major players in the additive manufacturing market have carved out their niches in terms of furthering the design, materials, print parameters, operational requirements, control systems, pre-processing and post-processing stations, and output characteristics of their machines [4]. The ISO/ASTM bodies have established standards for 3D printing processes, with the terminologies that are most widely used globally based on the *ISO/ASTM 52900 Standard Terminology for Additive Manufacturing—General Principles—Terminology* which became the *ASTM 52900:2015 Standard Terminology for Additive Manufacturing Technologies* in 2020 [5]. The standard set of 3D printing technologies is defined under this standard which includes the above list of seven processes. For adherence to and maintaining the standards for consistency in the published literature, table 2.1 highlights the possible techniques under the generalized processes mentioned above.

2.1.2 Comparison of key distinctions in 3D printing techniques

The above techniques require an understanding of the critical process requirements, including technology, pre-processing of the object using simulation, object orientation, print parameter settings, heating/quenching effects, support structure positioning and removals, and post-processing (sanding, polishing, painting, aesthetics, etc) [6, 7]. Table 2.2 gives details of the generalized printing processes and their basic differences, and their suitability for different application.

2.1.3 Brief descriptions of 3D printing techniques

The 3D printing process in general is simply the addition of material layers as opposed to the machining of subtractive manufacturing processes. To fabricate a physical object, significant variations in the commercial processes arise from the technology utilized, the instructional stages, printing parameters, pre- and post-processing methodologies, materials, accuracy, and the geometry of the model. Let us examine each of the processes in detail to understand their applications [8].

Table 2.1. Generalized and commercial terms for various 3D printing processes.

Generalized 3DP process	Commercial terminology
Binder jetting	Sand binder jetting
	Metal binder jetting
	Plastic binder jetting
Direct energy deposition	Electron beam additive manufacturing (EBAM)
	Laser engineered net shaping (LENS)
	Direct metal deposition (DMD)
	Wire arc additive manufacturing (WAAM)
	Rapid plasma deposition (RPD)
Material extrusion	Fused deposition modelling (FDM)
	Fused filament fabrication (FFF)
Material jetting	Material jetting (MJ)
	Drop on demand (DOD)
	Nanoparticle jetting (NPJ)
	Colour jet printing (CJP)
Powder bed fusion	Selective laser sintering (SLS)
	Direct metal laser sintering (DMLS)/Selective laser melting (SLM)
	Electron beam melting (EBM)
	Multi jet fusion (MJF)
Sheet lamination	Laminated object manufacturing (LOM)
	Ultrasonic consolidation (UC)
	Selective lamination composite object manufacturing (SLCOM)
	Computer-aided manufacturing of laminated engineering materials (CAM-LEM)
	Selective deposition lamination (SDL)
	Composite based additive manufacturing (CBAM)
Vat polymerization	Stereolithography (SLA)
	Digital light processing (DLP)
	Masked stereolithography (MSLA)
	Programmable photopolymerization (P3)
	High area rapid printing (HARP)
	Lithography-based metal manufacturing (LMM)
	Light enabled additive production (LEAP)
	Microstereolithography (μSLA)/Projection microstereolithography (PμSL)
	Two-photon polymerization (2PP or TPP)

2.1.3.1 Binder jetting

The binder jetting 3D printing process requires a liquid bonding agent to selectively bind regions of a bed of powdered material. As shown in figure 2.1, an initial layer of powder on the build platform is a mandatory requirement. A print head moves over

Table 2.2. Comparison of the key techniques of generalized 3D printing processes.

3D printing process	Advantages	Disadvantages	Possible materials	Dimensional accuracy	Applications
Binder jetting	Low cost, large build volumes, vivid colour reproduction, fast print speeds, support-free design	Mechanical properties not at par with parts printed using the powder bed fusion process	Sand, polymers, metal powder such as stainless steel or bronze, full-colour sand, silica for sand casting, and ceramic–metal composites	±0.2 mm (metal) or ±0.3 mm (sand)	Metal parts for functional usage, full-colour models, and sand casting
Direct energy deposition	Support structures rarely needed, composite metals, three-dimensional operation	Expensive, poor surface texture and finish with mandatory post-processing	Metals such as wires and powders	±0.1 mm	Automotive and aerospace components, functional prototypes, and end-use parts
Material extrusion	Inexpensive or rarely expensive, a wide range of materials, a wide range of applications	Not suitable for functional and strength intensive applications	Filaments of plastic such as PLA, ABS, PC, PC-ABS, PET, PETG, TPU, nylon, ASA, HIPS, carbon fibre, and many more	±0.5 mm	Prototypes, electrical housings, form and fit parts jigs, fixtures, patterns for investment casting, etc
Material jetting	Full-colour palette, multi-material possibility, best surface finish	More expensive than vat polymerization, brittle outputs, not suitable for mechanical parts	Resins that show photopolymerization (transparent, translucent, castable, dental, medical, high temperature, high strength, etc)	±0.1 mm	Medical models, injection moulds for low-runs, full-colour product prototypes, etc

Technology	Advantages	Disadvantages	Materials	Accuracy	Applications
Powder bed fusion	Functional parts with complex geometries and excellent mechanical properties	Slow and long print time, requirement for post-processing, thermal distortion	Powders of thermoplastic materials such as Nylon 6, Nylon 11, Nylon 12, etc, metals such as steel, titanium, aluminium, cobalt, etc, and ceramics	±0.3 mm	Functional parts, part production under low-run and complex parts such as ducting, venting, and flow pipes
Sheet lamination	Composite parts, quick print time, inexpensive	High wastage post-print and high levels of post-processing	Paper, polymers, and metals in the form of sheets	±0.1 mm	Non-functional prototypes, multi-colour prints, and casting moulds
Vat polymerization	Smooth surface finish, fine and exquisite details in the prints, high accuracy applications like smart sensors, and microfluidic devices	Expiry of resin, limited material availability, post-curing of parts, and post-processing (removal of resin using reagents)	Photopolymer resins are castable, transparent, biocompatible, etc	±0.15 mm	Jewellery casting, dental moulds, and casts, injection mould polymer prototypes

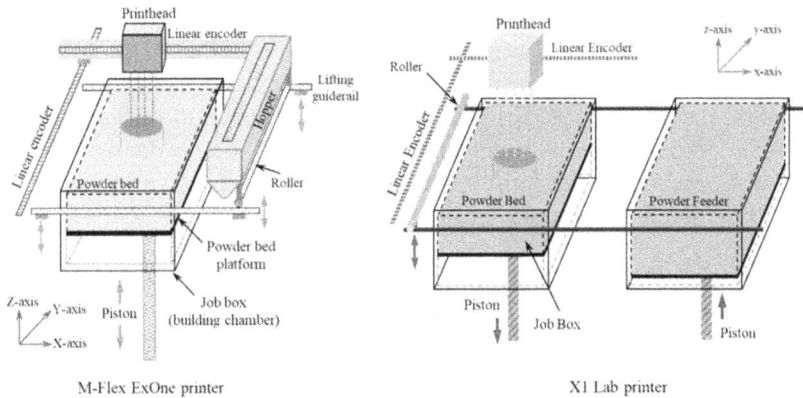

Figure 2.1. Binder jetting process. (Reproduced with permission from [9]. Copyright 2021 Elsevier. CC-BY 4.0.)

the powder bed and deposits droplets of binder fluid, typically 80 microns in diameter. These binder droplets in turn bind the powder particles together, producing each layer of the part. After each layer, the powder bed is lowered for a new layer of powder to be spread, on which the print head then deposits the binder. This process continues until the complete part is printed. The printed part is left in the powder bed for the binder to cure to gain strength, after which the vat is removed, and any unbound powder is cleared by blowing compressed air or is sucked up using a vacuum duct [9].

2.1.3.1.1 Sand binder jetting
As the name indicates, in sand binder jetting the bed is filled with fine sand particles and the binder is used to to produce parts by binding the sand. Parts of sandstone or gypsum can be made using this technique. The cores and moulds are removed from the build area after the printing process and cleaned to remove loose sand particles. After this, the moulds are usually ready for casting. After casting the mould is broken and the final metal component is removed. The sand binder jetting process enables foundries to scale up using 3D-printed moulds and supports faster turn-around times. This enables the possibility of rapid design changes, and exceptional design freedom to consolidate assemblies and create unique geometries. The most important advantage of this process is that it can be integrated into existing manufacturing or foundry processes without any disruption [10].

2.1.3.1.2 Metal binder jetting
Printing metal parts using the binder jetting process enables the production of complex parts beyond conventional manufacturing capabilities. The difference to the other binder jetting processes is that the material bed consists of metal powder and the binder is a polymer. Functional metal parts are made using secondary processes such as sintering or infiltration after the binder jetting process. This enables the parts to become functional and provides the requisite mechanical properties. In the sintering process, the green part of the binder jetting process is

cured inside an oven, followed by sintering in a furnace, reaching a density of ~97%. Non-uniform shrinkage needs to be addressed in the design stage to achieve the required dimensional accuracy. In the infiltration process, after curing inside a furnace where the density reaches ~60%, bronze is infiltrated into the voids in the part through capillary action, resulting in an object of ~90% density [11]. Objects made using the metal binder jetting process do, however, have lower mechanical properties than metal parts made using powder bed fusion.

2.1.3.1.3 Plastic binder jetting

The plastic binder jetting process has a similar methodology to metal binder jetting except for the fact that it does not require a furnace-based sintering step. It involves a base material of plastic powder and a liquid binding agent. After printing, the parts are used directly after cleaning without any further post-processing. However, the part can be infilled with another material for more strength (due to the added density). This step would need to be followed by curing, polishing, and painting. Binder jetting with polymers can produce parts at a lower cost than powder bed fusion due to faster print speeds and lower complexity [12].

2.1.3.2 Direct energy deposition

The direct energy deposition process is one of the most popular process choices for fabricating smooth metal parts and is used significantly to repair objects, particularly cracks and crevices in metals. As shown in figure 2.2, the material is fed in and fused using powerful thermal energy as it is deposited. The source of energy could be an electron beam, laser, or plasma. The material is fed as wires or powder to the heat source which melts it as it leaves the nozzle to form complex geometries and shapes. For the powder form of input, there is a need for the material to be sprayed along with an inert gas to reduce or eliminate the possibility of oxidation. This method also allows the mixing of several metal powders (composite) to achieve varied results and can print in all three dimensions, which allow for complex parts to be printed and repaired. The cost of the inclusion of the inert gas and the high amount of unused powder which goes off-target and remains unmelted makes this technique less

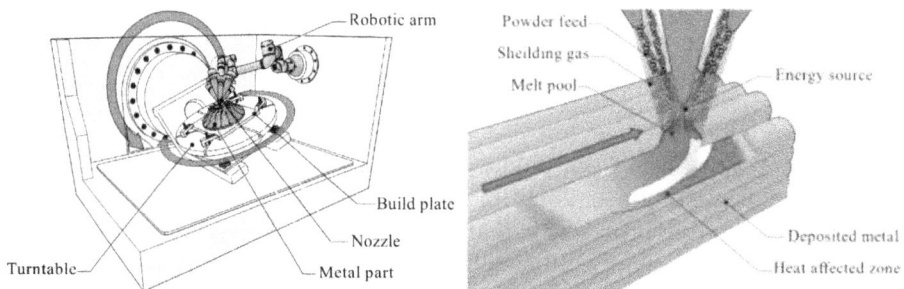

Figure 2.2. The direct energy deposition process. (Reproduced with permission from [14]. Copyright 2022 Elsevier. CC-BY 4.0.)

preferred compared to the powder bed fusion processes. This technique also requires considerable post-processing [13].

2.1.3.2.1 Electron beam additive manufacturing (EBAM)

In the EBEAM technique, both wire and powder forms of material feed can be used. The EBEAM requires a vacuum environment to reduce the possibility of contaminants in the final product output, and therefore does not have a need for inert gas flow. The electron beam creates a melt pool and adds material layer by layer. Commonly used metals include copper, cobalt, nickel, titanium, and tantalum. This technique is mostly used for fabricating medical implants and thus alloys of titanium are the most widely used [15].

2.1.3.2.2 Laser engineered net shaping (LENS)

In this technique, metal powder is fed through one or more nozzles and fused via a powerful laser inside an insulated chamber. The object is built layer-by-layer as the nozzle and layer move around in a three-dimensional manner. The insulation is against oxygen and moisture to prevent any contamination and thus to produce a clean part. An inert gas, usually argon, is used for the purpose of insulation. The commonly used metals include aluminium, copper, stainless steel, and titanium. This method is preferred for the repair of high-end aerospace and automotive components such as jet blades, pistons casing, etc. Post-processing is required after the printing to ensure a smooth surface finish [16].

2.1.3.2.3 Direct metal deposition (DMD)

The DMD method is the only metal 3D printing method that does not use a powder bed. This involves the transformation of powdered metal into a solid object wherein the powder is fed through a nozzle and propelled into a laser beam. Using a layer-by-layer strategy, the printer head, composed of the laser beam and the feed nozzle, can scan the substrate to deposit successive layers. This metal additive manufacturing technique is also called laser metal deposition (LMD). This approach uses all types of metal particles, from basic ones such as steel and aluminium to more technical ones such as cobalt, copper, and titanium. Custom alloys are also possible. This technique is reliable and stable in producing aerospace and automotive end-use parts [17].

2.1.3.2.4 Wire arc additive manufacturing (WAAM)

In this method an arc welding process is used to print metal parts. It works by melting a metal wire using an electric arc as the heat source. The process is usually controlled using a robotic arm and the wire is extruded in the form of a bead on the substrate. The materials used include a wide range of alloys from stainless steel and aluminium alloys to nickel-based and titanium alloys. The WAAM technique is one of the lesser-known metal 3D printing technologies, but has enormous potential for large-scale applications across multiple industries. The melting of the wire and the lack of need for a vacuum chamber render this process easier and less expensive when compared to the PBF process. The parts produced by WAAM have a high density and strong mechanical properties [18].

2.1.3.2.5 Rapid plasma deposition (RPD)

This is a commercial additive manufacturing process that delivers aerospace-grade parts with reduced lead time and cost. Structural titanium parts can be manufactured using this process wherein high-volume manufacturing can be done. The process involves a titanium wire, argon gas for inert flow, and a plasma arc. This is a controlled plasma arc additive manufacturing method. This technique can produce titanium parts 50–100 times faster than the conventional PBF process with a 25%–75% lower material requirement when compared to forging. This process is suitable for applications across industries, including aviation, space, transportation, oil, and gas, and maritime [19].

2.1.3.3 Material extrusion

As the name suggests, the material is extruded out of a heated nozzle onto a platform. The material is usually plastic and is pushed through a heated nozzle which deposits the material on a build platform along a predetermined path, as shown in figure 2.3. The path is determined by an algorithm that allows the filament sufficient time to cool and solidify before the next layer is laid out. This technique allows the extrusion of metal paste, bio-gels, concrete, chocolate, and a wide range of other materials, but plastics are the most common [20].

2.1.3.3.1 Fused deposition modelling (FDM)/fused filament fabrication (FFF)

This is one of the most common and affordable 3D printing techniques globally. A spool of filament is loaded into the 3D printer and laid onto the build platform through a printer nozzle in the extrusion head. The nozzle is heated to the desired temperature based on the material that needs to be extruded and a motor pushes the filament through the nozzle thereby causing it to melt. The printer moves the extrusion head across the build platform to specified coordinates to deposit the molten material. After each layer, the build platform moves down to allow for the next layer to be laid down. This process of printing cross-sections is repeated, building layer upon layer until the object is fully formed. Support structures are crucial in the design when the extrusion requires supports, such as for an overhang or hidden crevice. FDM is used is for 3D printing of buildings by extruding concrete

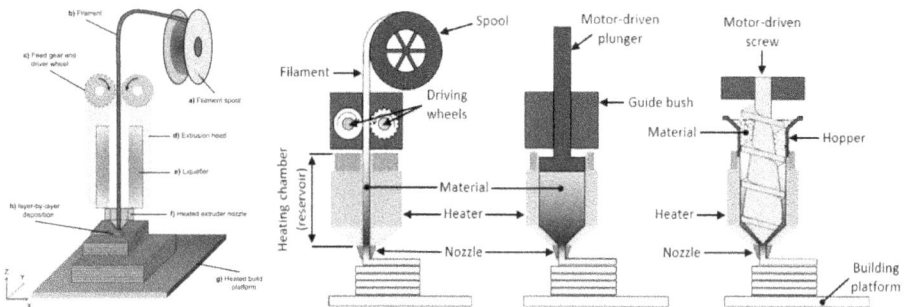

Figure 2.3. Material extrusion process. (Reproduced with permission from [21, 22]. Copyright 2021 Elsevier. CC-BY 4.0.)

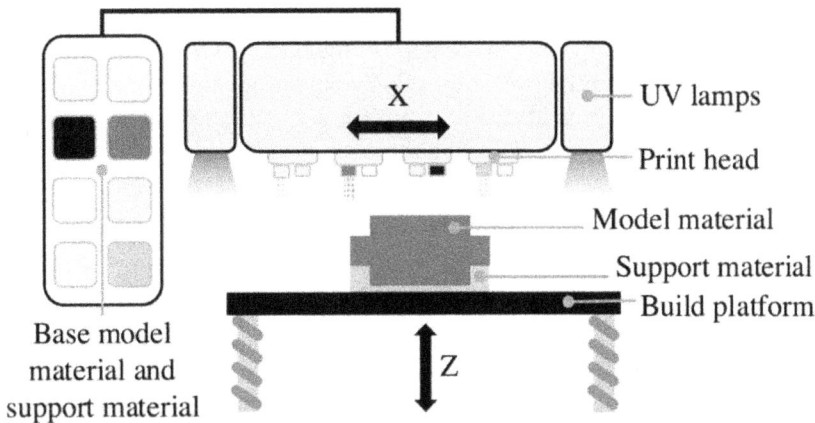

Figure 2.4. The material jetting process. (Reproduced with permission from [26]. Copyright 2022 Elsevier.)

or clay, food by extruding chocolate, and organs by extruding live cells or bio-gels, and suggests the possibility that anything that can be extruded, can be printed [23, 24].

2.1.3.4 Material jetting processes

Material jetting is a 3D printing process where droplets of material are selectively deposited and cured on a build plate. As shown in figure 2.4, the materials are photopolymers or wax droplets that can be cured upon exposure to light. The objects are built up one layer at a time to form the final part. The technique allows varied materials to be printed at the same time, allowing for the fabrication of parts in varied colours and textures suitable for applications in medical science, for surgery preparation, or the automotive industry, such as car models, etc [25].

2.1.3.4.1 Material jetting (MJ)

The process is similar to a standard inkjet printer, except that multiple layers are built upon each other to create a solid part as opposed to a single layer of ink. The print head jets hundreds of tiny droplets of photopolymers which are then cured/solidified by ultraviolet light. After each layer is cured, the build platform is lowered for the addition of the next layer. In comparison to the point-wise deposition techniques in other processes, MJ machines deposit the material in a rapid, line-wise fashion which allows for the fabrication of multiple objects in a single line with no effect on the build speed. This technique can build parts faster than other techniques if parts are arranged with optimized space between each build line. The supports are built using dissolvable material which can be removed by curing them inside a support removal basket with dissolving agents. MJ is one of the only types of 3D printing technology to offer objects made from multiple materials with full-colour reproduction [26].

2.1.3.4.2 Drop on demand (DOD)

This technology uses a pair of inkjets to deposit the build material and the dissolvable support material. The material is usually wax-like in nature and the print head follows a predetermined path to jet material in a precise manner creating the cross-sectional area of an object. A fly cutter is usually used to skim the build area to make it flat for the next layer. This technique is preferred for creating patterns for lost-wax casting or investment casting and mould-making applications [27].

2.1.3.4.3 Nanoparticle jetting (NPJ)

This technique enables the fabrication of parts made of metals and ceramics. This technique involves the jetting of liquid suspensions via a dispersion technology wherein the suspensions contain solid nanoparticles of build and support materials. The liquid suspensions are the basis of this process. The parts need to be post-processed, i.e. the support material needs to be removed using removal agents. The outputs have elevated levels of detail, finish, and accuracy, while allowing the possibility of delivering complex physical geometries [28].

2.1.3.4.4 Colour jet printing (CJP)

This technique involves two major components, namely the core material and the binder. The core material is spread on top of the build platform as thin layers and the binder is jetted selectively from inkjet print heads helping the core material to solidify. The build platform is lowered for subsequent layers to be spread and printed. This method allows for the printing of parts in full colour. This method is well suited for a field such as architecture, as demonstration and concept models can be printed three-dimensionally [29].

2.1.3.5 Powder bed fusion

In the powder bed fusion process, a thermal energy source selectively induces fusion between powder particles made of plastic, metal, or ceramic. This occurs inside a build area to create a solid object (figure 2.5). The printer spreads a thin layer of powdered material on the print bed using a wiper or blade which is followed by the energy source that fuses the powder at specific points. This process is repeated until

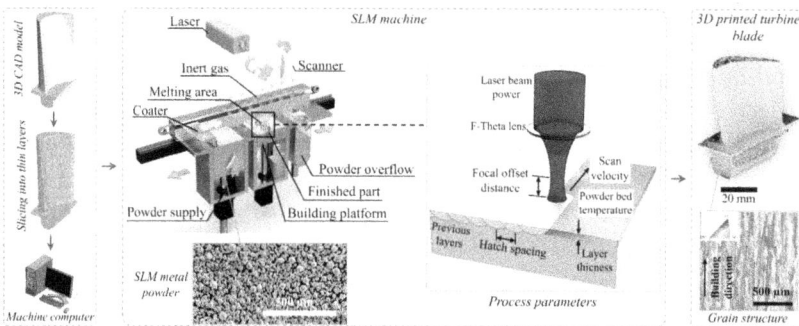

Figure 2.5. The powder bed fusion process. (Reproduced with permission from [12]. Copyright 2022 Elsevier.)

the entire object is formed. The part is supported and encased inside the unfused powder [30].

2.1.3.5.1 Selective laser sintering (SLS)

A combination of polymer powder and the powder bed fusion technique is known as selective laser sintering. In this method a vat/bed of polymer powder is heated below its melting point, and a wiper or recoating blade deposits a thin layer (–0.1 mm) onto a build platform. A laser (CO_2 or fibre) scans the surface and selectively sinters the powder to solidify across its cross-section. The focusing of the laser is achieved by a pair of automated galvos. After each cross-section is scanned, the build platform is lowered for the next coat of the powder, followed by laser sintering. These steps are repeated until all the objects are complete. The powder that has not been sintered remains in place to support the object, which reduces or eliminates the need for support structures [31].

2.1.3.5.2 Direct metal laser sintering (DMLS)/selective laser melting (SLM)

Both these techniques are used to produce metal parts like SLS. DMLS does not melt the powders but heats them to a point to fuse them at a molecular level. SLM enables the fusion of powder to form a homogeneous part [32]. Therefore, DMLS is used where there is a requirement for alloys and SLM is used when pure metal parts with a single melting temperature are required (e.g. titanium). Unlike SLS, the DMLS and SLM processes require structural support to limit the possibility of any distortion (even though the surrounding powder provides physical support). The parts produce must be cleaned of remaining powder and then post-processed (heat-treated, machined, and polished) to obtain the final part. After printing, parts are also typically heat-treated to relieve any residual stresses in the parts owing to warping that may be produced due to elevated temperatures. Parts produced using this method are highly suitable for direct end-use applications across various domains, including aerospace, automotive, renewable energy, etc.

2.1.3.5.3 Electron beam melting (EBM)

Distinct from other PBF techniques, EBM uses a high-energy electron beam to induce fusion between metal particles. The beam scans across a thin layer of powder causing localized melting and solidification of the part across its cross-section. This occurs layer-by-layer to create a solid object. The EBM technique can only be applied using conductive metals such as copper and requires a vacuum chamber to avoid contamination and to maintain a controlled environment. The EBM technique has a superior build speed owing to the higher energy density of the beam. However, properties such as minimum feature size, powder particle size, layer thickness, and surface finish are typically larger [33].

2.1.3.5.4 Multi-jet fusion (MJF)

This technology come under PBF although the name suggests its similarity to the binder jetting technique. It employs multiple ink-jet heads, cost-effective mechanics, and the capability to jet various materials, all based on principles of material science. The printer lays down a layer of material powder followed by an inkjet head that deposits a fusing and detailing agent onto it. An infrared heating unit then moves across the print making the fusing agent jetted layers melt together whereas the detailing agent jetted layers will stay as powder. Since there are individual inkjet heads for material coating and agent distribution, there is better control of optimization. This technique, just like SLS, eliminates the need for modelling supports as the powder layers support the parts printed above them. This technique is also unique compared to other SLS methods because each new material and agent layer is deposited while the previous layer is still in its molten state. This lets both layers fuse completely, delivering improved print durability and finer detail. The entire powder bed is shifted to a separate station for post-processing wherein the unused powder is vacuumed up to be reused [34].

2.1.3.6 Sheet lamination

This technique involves stacking and laminating sheets of very thin material together to produce a 3D object, as shown in figure 2.6. The fusion of the sheets can be done using a variety of methods such as heat, sound, and glue, depending on the material, such as paper, polymers, and metals. This is one of the less accurate 3D printing technologies and the printed part requires a lot of processing. Laser cutters and computer numerical control (CNC) machines are used during the printing process to form parts with desired shapes. The outputs are not strong and mainly serve as

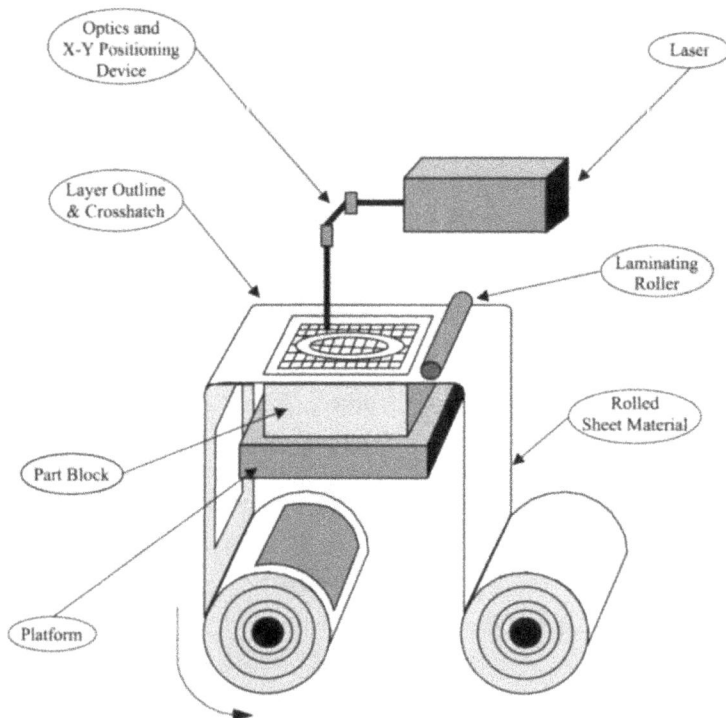

Figure 2.6. The sheet lamination process. (Reproduced with permission from [35]. Copyright 2022 Elsevier.)

aesthetic works. This technique is used to produce cost-effective, non-functional prototypes at a high speed. It can also produce composite items, as the materials used can be swapped around during the printing process [35].

2.1.3.6.1 Laminated object manufacturing (LOM)

LOM is the most common sheet lamination technique wherein sheets of paper are layered on top of each other and glued together to form a bond. Like many other techniques, layers are built one at a time, but just the bonding alone cannot form the object. A knife, laser, or CNC router is used to cut the shape out of the layered object. A larger amount of glue is applied to the portions which will form part of the final object and a smaller amount is applied everywhere else. The cutter cuts a 2D cross-section of the final print as layers are added simultaneously. Prints are quick, affordable, and suitable for large objects. However, the prints are not strong and lose their properties as they age and require post-processing such as drilling and machining, producing excessive waste compared to all other techniques [36].

2.1.3.6.2 Ultrasonic consolidation (UC)

UC or ultrasonic additive manufacturing (UAM) uses ultrasonic vibrations and pressure using a sonotrode to fuse thin sheets of metal at low temperatures. Because of this, the metal sheets are bonded together instead of melting and fusing. This happens due to the breakdown of oxides on the surface of the sheet. This technique allows for the mixing of multiple metals owing to the lower temperature of operation and the avoidance of actual inter-mixing. Like all other sheet lamination techniques, here a CNC router is required to cut out the cross-section of the desired shape after every layer. Due to the cutting process, there is more waste with this method than with other metal 3D printing approaches, and the cutter can also be used to produce details and designs as the print is being formed. Post-processing, such as finishing and polishing, is required but the high speed of operation and lower temperatures gives this method its unique advantages [37].

2.1.3.6.3 Selective lamination composite object manufacturing (SLCOM)

This is one of the latest methods of sheet laminated fabrication wherein the sheets are made of laminated thermoplastic composite fabric. This is one of the very few techniques wherein custom-made thermoplastic reinforced unidirectional or multi-directional woven fibres can be laminated together. The composites can be tailored for properties such as toughness, environmental resistance, high wear resistance, low flammability, etc. As with all other sheet lamination methods, they are glued together, and a CNC router is used to cut out the 2D cross-section of the final object [38].

2.1.3.6.4 Computer-aided manufacturing of laminated engineering materials (CAM-LEM)

Advanced ceramic materials such as alumina, silicon carbide, and silicon nitride have found applications in gas turbine engines but have a limited scope of manufacturability owing to limitations such as temperature, density, and material complexities. The CAM-LEM technique allows for powdered material to be fused and confined to sheets of a particular thickness. These sheets are then cut as per the cross-sections required in the geometry. The assembly involves the fixing of the position of the sheets relative to a pre-existing stack and thereby achieves intimate interlayer contact. The assembled stack is then sintered for high-integrity bonding. The laminated green object, which is considered monolithic, is then fired to densify and allow for microstructural evolution. The parts coming out of this technique are very functional in nature with exact geometric form [39].

2.1.3.6.5 Selective deposition lamination (SDL)
This process is similar to the LOM technique, wherein layers of adhesive-coated paper (or plastic or metal laminates) are successively glued together with a heated roller and cut to shape with a laser cutter layer by layer. A roller brings in the next sheet and lays it over the last, and the process is repeated. Very low-cost materials, such as off-the-shelf copier paper, can be used as the input material and the finishing can be done using wood finishing techniques, since the object produced have similar characteristics to wood. This approach differs from LOM, most notably in the gluing process. In SDL only the parts that will make up the object are glued, whereas in LOM the entire sheet is glued uniformly [40].

2.1.3.6.6 Composite-based additive manufacturing (CBAM)
This process is specifically designed for composites. The material possibilities include carbon fibre and fibreglass paired with nylon and polyether ether ketone (PEEK). This process allows for parts to be produced that are stronger than those from other additive manufacturing technologies, and is faster than traditional composite manufacturing methods involving better design freedom and complexity [41].

2.1.3.7 Vat polymerization
This technique involves the selective curing of a photopolymer resin inside a vat (figure 2.7). Light is directed precisely to a specific point on a thin layer of liquid plastic. Three major forms of vat polymerization techniques are stereolithography (SLA), digital light processing (DLP), and masked stereolithography (MSLA). The fundamental difference between these types of 3D printing technology is the light source they use to cure the resin [42]. The details of each method are described below along with some newer polymerization methods.

2.1.3.7.1 Stereolithography (SLA)
SLA holds historical significance as the world's first 3D printing technology. Invented by Charles Hull in 1986 [44], he patented and founded 3D Systems to commercialize it. SLA technology uses mirrors which are called galvanometers positioned on the x- and y-axes which aim a laser beam on a vat of resin selectively curing and solidifying a cross-section of the object inside this building area, building it up layer by layer [45]. Most SLA printers used solid-state lasers which makes the process slow owing to the point-based tracing of a cross-section area. This led to the invention of DLP, which hardens an entire layer.

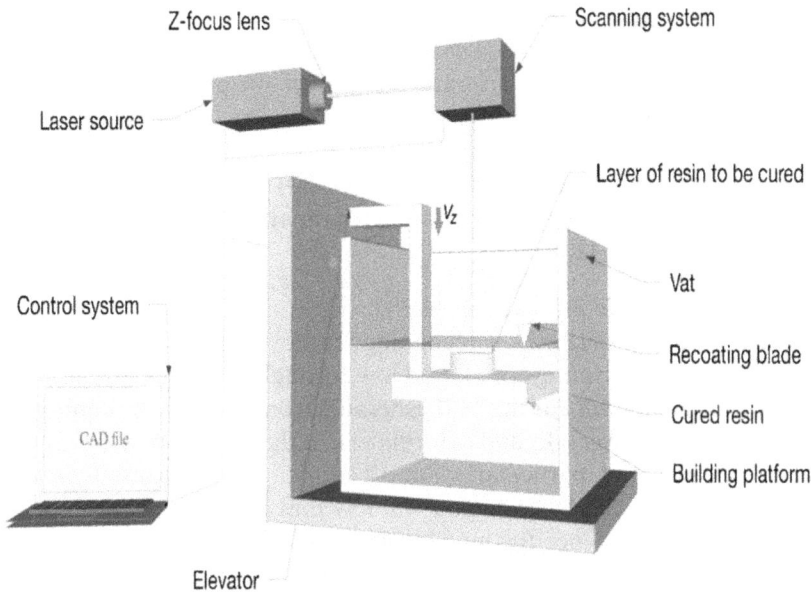

Figure 2.7. Vat photopolymerization process. (Reproduced with permission from [43]. Copyright 2020 Elsevier.)

2.1.3.7.2 Digital light processing (DLP)

The key difference of this method in comparison to SLA is that DLP uses a digital light projector to flash a cross-sectional image of an entire layer at once, or multiple flashes for several parts or a larger part. The image is composed of several pixels which are square in shape, resulting in a layer formed from rectangular blocks called voxels. This is because the projector is a digital screen. The light could be a UV light source or a light emitting diode (LED) that is directed in to the vat using a digital micromirror device (DMD) which is an array of micromirrors that controls the location of the light projection and generates the light pattern for the build surface. The products printed using this technique can be used in a variety of applications in jewellery, dentistry, and microfluidics, such as energy harvesting devices and smart sensors [46].

2.1.3.7.3 Masked stereolithography (MSLA)

This technique is like the SLA technique with the addition of an LCD screen which acts as a mask. Like DLP, the LCD screen is a digital display composed of square pixels. The pixel size is responsible for the granularity of the print thus fixing the x–y accuracy, unlike DLP wherein the zooming/scaling of the lens changes the granularity. The MSLA technique uses an array of hundreds of individual emitters rather than a single-point light source such as a laser diode or DLP bulb. The low cost of LCD units has made this technology suitable for the budget desktop resin printing sector [47].

2.1.3.7.4 Programmable photopolymerization (P3)

This technique is a patented process of Stratasys, wherein the print process is tightly synchronized, including pneumatic controls to reduce the pull forces during the print process, resulting in exceptional surface quality without sacrificing speed or isotropy. Material options include hot temperature, tough, general purpose elastomers, and specialty materials with electrostatic discharge (ESD) properties, ceramic attributes, and castability. Applications range from biocompatible medical devices to aerospace-grade components [48].

2.1.3.7.5 High area rapid printing (HARP)

This technique is based on the SLA technology but overcomes its limitations, which are limits to size, slow printing speed, and low throughput for a large print volume. The only downside of HARP is the heat generated from the system running at higher speeds, which may cause cracks and deformation in the printed parts. This technique prints in a liquid polymer, particularly liquid Teflon to avoid the disadvantages. This process uses a projected light through a window that solidifies the resin on top of a vertically moving plate. Since the material is non-sticky in nature, it can flow over the window to reduce heat and circulated through a cooling unit during high print speeds. The non-sticky nature also helps avoid the cleaving of the bottom layer from the vat as it resists adhesion to the print bed [23].

2.1.3.7.6 Lithography-based metal manufacturing (LMM)

LMM is an additive manufacturing technology that uses the process of photopolymerization for producing metal models and production parts. Here the metal powder is homogeneously dispersed in a light-sensitive resin and selectively polymerized upon exposure to light. It can produce parts with the same material properties as metal injection moulding (MIM). The self-supporting function of the vat/bed allows for the volume-optimized placement of different geometries on a single building platform, without the need to add support structures. The resolution of the parts is exceedingly high with the possibility of intricate design and features with smooth surface aesthetics [49].

2.1.3.7.7 Light enabled additive production (LEAP)

This technique is an advancement of DLP wherein the focus is to achieve high printing speeds of parts with production-grade performance. This technique allows for custom volume production. It adds a layer of interface between the optical engine and the vat containing the liquid resin/photopolymer. This enables quicker cross-sectional printing of parts thereby enabling applications in the dental, footwear, eyewear, medical, and manufacturing domains [50].

2.1.3.7.8 Microstereolithography (μSLA)/projection microstereolithography (PμSL)

This process involves exposing photosensitive material (liquid resin) to an ultraviolet laser. The general process is the same, which involves pouring resin into a tank, lowering of the build platform into the resin, a laser to draw a cross-section of the 3D part layer by layer, while the platform is lowered further into the chamber. The difference is the sophistication of the lasers and the addition of lenses, which can generate unbelievably small points of light, and specialized resins. The PμSL process is similar to μSLA, except that instead of a laser, PμSL uses ultraviolet light from a projector. The technique allows for rapid photopolymerization of an entire layer of liquid polymer using a flash of UV light at the micro-scale resolution, so it is significantly faster [51].

2.1.3.7.9 Two-photon polymerization (2PP or TPP)
This technology deploys a pulsed femtosecond laser to trace 3D patterns at depth in a vat filled with special photosensitive resin. The light photons which are generated from the laser are absorbed by the resin at such a level that resolutions of less than 1 μm are achievable. Therefore, this technique falls under the category of nanofabrication technology. This resolution is one of the highest accuracies among micro-3D printing solutions. It will be beneficial for medical innovations including tissue engineering, medical implants, micromechanics, smart sensors, microfluidic devices, and energy harvesting devices. The downside is that the technology and materials are expensive, and the printers can be slower than other technologies [52].

2.2 Conclusion

The utilization of 3D printing in the domain of MEMS and microfluidic devices has been low considering the advantages of this additive technology, which allows for tailoring 3D shapes in every desired way, making it reliable and process friendly. The advent of customized 3D printing techniques has reduced the challenges in printing MEMS devices, such as minimum feature size, high resolution, thermal shrinkage, misalignments, etc. However, it is still to be seen if a commercial methodology which is affordable and user friendly in terms of the printing process, parameter selection, and output quality can result in a standardized manufacturing technique for MEMS devices. The 3D printing of MEMS in such a standardized manner could result in the benefits of faster production and greater affordability when printed in higher numbers.

References

[1] Hajare D M and Gajbhiye T S 2022 Additive manufacturing (3D printing): recent progress on advancement of materials and challenges *Mater. Today Proc.* **58** 736–43
[2] Goel S, Venkateswaran P S, Prajesh R and Agarwal A 2015 Rapid and automated measurement of biofuel blending using a microfluidic viscometer *Fuel* **139** 213–9
[3] González-Henríquez C M, Sarabia-Vallejos M A and Rodriguez-Hernandez J 2019 Polymers for additive manufacturing and 4D-printing: materials, methodologies, and biomedical applications *Prog. Polym. Sci.* **94** 57–116
[4] Siacor F D C *et al* 2021 On the additive manufacturing (3D printing) of viscoelastic materials and flow behavior: from composites to food manufacturing *Addit. Manuf.* **45** 102043
[5] Alexander A E, Wake N, Chepelev L, Brantner P, Ryan J and Wang K C 2021 A guideline for 3D printing terminology in biomedical research utilizing ISO/ASTM standards *3D Print. Med.* **7** 8
[6] Ranjan R, Kumar D, Kundu M and Chandra Moi S 2022 A critical review on classification of materials used in 3D printing process *Mater. Today Proc.* **61** 43–9

[7] Venkateswaran P S, Kashyap D, Agarwal A and Goel S 2015 Computational analysis of a microfluidic viscometer and its application in the rapid and automated measurement of biodiesel blending under pressure driven flow *J. Comput. Theor. Nanosci.* **12** 2311–7

[8] Liacouras P and Wake N 2022 3D printing principles and technologies *3D Printing for the Radiologist* ed N Wake (Amsterdam: Elsevier) ch 5 pp 61–73 https://sciencedirect.com/science/article/pii/B9780323775731000166

[9] Mostafaei A *et al* 2021 Binder jet 3D printing—process parameters, materials, properties, modeling, and challenges *Prog. Mater. Sci.* **119** 100707

[10] Sivarupan T *et al* 2021 A review on the progress and challenges of binder jet 3D printing of sand moulds for advanced casting *Addit. Manuf.* **40** 101889

[11] Lu S L, Meenashisundaram G K, Wang P, Nai S M L and Wei J 2020 The combined influence of elevated pre-sintering and subsequent bronze infiltration on the microstructures and mechanical properties of 420 stainless steel additively manufactured via binder jet printing *Addit. Manuf.* **34** 101266

[12] Park S, Shou W, Makatura L, Matusik W and Fu K 2022 3D printing of polymer composites: materials, processes, and applications *Matter* **5** 43–76

[13] Aprilia A, Wu N and Zhou W 2022 Repair and restoration of engineering components by laser directed energy deposition *Mater. Today Proc.* **70** S1

[14] Armstrong M, Mehrabi H and Naveed N 2022 An overview of modern metal additive manufacturing technology *J. Manuf. Process* **84** 1001–29

[15] Jia Y, Mehta S T, Li R, Rahman Chowdhury M A, Horn T and Xu C 2021 Additive manufacturing of ZrB2–ZrSi2 ultra-high temperature ceramic composites using an electron beam melting process *Ceram. Int.* **47** 2397–405

[16] Hamilton J D *et al* 2022 Property-structure-process relationships in dissimilar material repair with directed energy deposition: repairing gray cast iron using stainless steel 316L *J. Manuf. Process* **81** 27–34

[17] Xu W 2022 Direct additive manufacturing techniques for metal parts: SLM, EBM, laser metal deposition *Encyclopedia of Materials: Metals and Alloys* ed F G Caballero (Oxford: Elsevier) pp 290–318

[18] Filippov A V *et al* 2021 Characterization of gradient CuAl–B4C composites additively manufactured using a combination of wire-feed and powder-bed electron beam deposition methods *J. Alloys Compd.* **859** 157824

[19] de Souza F M, Gupta R K, Yasin G and Nguyen T A 2022 Plasma produced by carbon nanotube-generated electron beam *Plasma at the Nanoscale* ed H Song, T A Nguyen, A Amrane, A Amine Assadi and G Yasin (Amsterdam: Elsevier) ch 2 pp 21–35

[20] Cano-Vicent A *et al* 2021 Fused deposition modelling: current status, methodology, applications and future prospects *Addit. Manuf.* **47** 102378

[21] Golab M, Massey S and Moultrie J 2022 How generalisable are material extrusion additive manufacturing parameter optimisation studies? A systematic review *Heliyon* **8** e11592

[22] Altıparmak S C, Yardley V A, Shi Z and Lin J 2022 Extrusion-based additive manufacturing technologies: state of the art and future perspectives *J. Manuf. Process* **83** 607–36

[23] Rocha D P *et al* 2023 Sensing materials: electrochemical sensors enabled by 3D printing *Encyclopedia of Sensors and Biosensors* 1st edn ed R Narayan (Oxford: Elsevier) pp 73–88

[24] Quero R F, Costa B M, de C, da Silva J A F and de Jesus D P 2022 Using multi-material fused deposition modeling (FDM) for one-step 3D printing of microfluidic capillary

electrophoresis with integrated electrodes for capacitively coupled contactless conductivity detection *Sens. Actuators* B **365** 131959

[25] Hayes B, Hainsworth T and MacCurdy R 2022 Liquid–solid co-printing of multi-material 3D fluidic devices via material jetting *Addit. Manuf.* **55** 102785

[26] Wei X, Zou N, Zeng L and Pei Z 2022 PolyJet 3D printing: predicting color by multilayer perceptron neural network *Ann. 3D Print. Med.* **5** 100049

[27] Davoodi E, Fayazfar H, Liravi F, Jabari E and Toyserkani E 2020 Drop-on-demand high-speed 3D printing of flexible milled carbon fiber/silicone composite sensors for wearable biomonitoring devices *Addit. Manuf.* **32** 101016

[28] Kunchala P and Kappagantula K 2018 3D printing high density ceramics using binder jetting with nanoparticle densifiers *Mater. Des.* **155** 443–50

[29] He Y *et al* 2022 Ink-jet 3D printing as a strategy for developing bespoke non-eluting biofilm resistant medical devices *Biomaterials* **281** 121350

[30] Awad A, Fina F, Goyanes A, Gaisford S and Basit A W 2021 Advances in powder bed fusion 3D printing in drug delivery and healthcare *Adv. Drug Deliv. Rev.* **174** 406–24

[31] Lekurwale S, Karanwad T and Banerjee S 2022 Selective laser sintering (SLS) of 3D printlets using a 3D printer comprised of IR/red-diode laser *Ann. 3D Print. Med.* **6** 100054

[32] Awad A, Fina F, Goyanes A, Gaisford S and Basit A W 2020 3D printing: principles and pharmaceutical applications of selective laser sintering *Int. J. Pharm.* **586** 119594

[33] Uçak N, Çiçek A and Aslantas K 2022 Machinability of 3D printed metallic materials fabricated by selective laser melting and electron beam melting: a review *J. Manuf. Process* **80** 414–57

[34] Persad J and Rocke S 2022 Multi-material 3D printed electronic assemblies: a review *Results Eng.* **16** 100730

[35] Srivastava M, Rathee S, Patel V, Kumar A and Koppad P G 2022 A review of various materials for additive manufacturing: recent trends and processing issues *J. Mater. Res. Technol.* **21** 2612–41

[36] Feygin M, Hsieh B and Melkanoff M A 1992 Laminated object manufacturing (Lom): a new tool in the cim world *Human Aspects in Computer Integrated Manufacturing* ed G J Olling and F Kimura (Amsterdam: Elsevier) pp 457–64

[37] Friel R J 2022 Metal sheet lamination—ultrasonic *Encyclopedia of Materials: Metals and Alloys* ed F G Caballero (Oxford: Elsevier) pp 134–40

[38] Sánchez del Río J, Pascual-González C, Martínez V, Jiménez J L and González C 2021 3D-printed resistive carbon-fiber-reinforced sensors for monitoring the resin frontal flow during composite manufacturing *Sens. Actuators* A **317** 112422

[39] Hofer A K, Kocjan A and Bermejo R 2022 High-strength lithography-based additive manufacturing of ceramic components with rapid sintering *Addit. Manuf.* **59** 103141

[40] Pilipović A 2022 Sheet lamination *Polymers for 3D Printing* ed J Izdebska-Podsiadły (Norwich, NY: William Andrew) ch 11 pp 127–36

[41] Bhatia A and Sehgal A K 2021 Additive manufacturing materials, methods and applications: a review *Mater. Today Proc.* **81**

[42] Davoudinejad A 2021 Vat photopolymerization methods in additive manufacturing *Additive Manufacturing* ed J Pou, A Riveiro and J P Davim (Amsterdam: Elsevier) ch 5 pp 159–81

[43] Zakeri S, Vippola M and Levänen E 2020 A comprehensive review of the photopolymerization of ceramic resins used in stereolithography *Addit. Manuf.* **35** 101177

[44] Hull C W 1984 Apparatus for production of three-dimensional objects by stereolithography *Patent Application* United States No 638905 (filed)

[45] Sankaran V P, Dubey S, Sharma A, Agarwal A and Goel S 2017 Stereolithographic 3D printed microfluidic viscometer for rapid detection of automobile fuel adulteration *Sens. Lett.* **15** 545–51

[46] Venkateswaran P S, Sharma A, Dubey S, Agarwal A and Goel S 2016 Rapid and automated measurement of milk adulteration using a 3D printed optofluidic microviscometer (OMV) *IEEE Sens. J.* **16** 3000–7

[47] Valizadeh I, Al Aboud A, Dörsam E and Weeger O 2021 Tailoring of functionally graded hyperelastic materials via grayscale mask stereolithography 3D printing *Addit. Manuf.* **47** 102108

[48] Rau D A, Forgiarini M and Williams C B 2021 Hybridizing direct ink write and mask-projection vat photopolymerization to enable additive manufacturing of high viscosity photopolymer resins *Addit. Manuf.* **42** 101996

[49] Borasi L *et al* 2021 3D metal freeform micromanufacturing *J. Manuf. Process* **68** 867–76

[50] Rouf S *et al* 2022 Additive manufacturing technologies: industrial and medical applications *Sustain. Oper. Comput.* **3** 258–74

[51] Troksa A L *et al* 2021 3D-printed nanoporous ceramics: tunable feedstock for direct ink write and projection microstereolithography *Mater. Des.* **198** 109337

[52] Bertsch A and Renaud P 2020 Microstereolithography *Three-Dimensional Microfabrication Using Two-Photon Polymerization* 2nd edn ed T Baldacchini (Norwich, NY: William Andrew) ch 1.2 pp 25–56

IOP Publishing

3D Printed Smart Sensors and Energy Harvesting Devices
Concepts, fabrication and applications
Sanket Goel and Sohan Dudala

Chapter 3

Additive and subtractive manufacturing

Srinivasa Prakash Regalla

In this chapter the relationship between traditional and modern subtractive manufacturing processes and additive manufacturing processes is investigated. Additive manufacturing is making rapid progress towards proving to be a viable option for industrial manufacturing in every sector. However, it is the judicious combination of additive manufacturing with traditional manufacturing processes that provides the optimum manufacturing cost and time. Subtractive manufacturing processes at the macro-scale are uneconomical for complex geometries but their machinery and process knowledge are well established and they have cheaper mass production costs. On the other hand, additive manufacturing is a near-net-shape process and geometric complexity does not pose a limitation. In recent years hybrid additive and subtractive processing has been gaining popularity. The processing and sustainability of these hybrid technologies are analysed and discussed in the last section.

3.1 Introduction

Manufacturing can be generally defined as the application of energy to convert raw materials into a useful product. The two entities involved in manufacturing, namely *materials* and *energy*, have limited availability in nature. Therefore, any manufacturing process that allows comparatively less consumption of these two entities may be treated as superior. Although manufacturing processes are as old as the Stone Age, their systematic development into documented and repeatable technologies happened more recently, in the last three centuries. The first notable developments occurred around 1750 during the first industrial revolution and the second major stage of development was during the early 1900s when the World Wars necessitated manufacturing technologies for the mass production of high quality artillery and weapons. The four major categories of traditional manufacturing technologies, namely, casting, forming, machining, and welding, underwent rapid and significant development in the first 50 years of the twentieth century.

3-1

Towards the end of the twentieth century, new materials began to arise on the horizon, promising special mechanical properties for the burgeoning industrial sectors of aviation and automobiles in the growth of globalized economies. Processing of the new materials with high strength and hardness to impart features with very small or intricate dimensions became a challenge and the existing four tracks of traditional manufacturing processes fell short. These new requirements led to the innovation of a new set of manufacturing processes, called non-traditional or unconventional manufacturing processes, in which the fundamental concept of the tool being mechanically harder than the workpiece has been nullified. The source and form of energy used to remove or to change the material has transcended the hitherto mechanical form and entered into chemical, electrical and electromagnetic forms. Materials of high brittleness such glass or ceramics could now be created with holes of diameters as small as a micron to depths of more than a hundred millimeters accurately with ultrasonic machining. Very high strength metal and alloy plates could now be cleaved like butter into slices with high accuracy and finish with wire electric-discharge machining. For sub-micron features of high complexity on the surfaces of metals and non-metals, electrochemical machining or chemical etching could be used.

The early part of the twenty-first century saw the revival of space explorations by more countries, and indigenous aerospace manufacturing expanded to newer countries such as India. Both consumer goods and high-technology products such as gas turbines and prosthetics benefited from leveraging the maturity reached by computer aided design technologies directly to produce products of higher geometric complexity and a lower number of components in shorter manufacturing cycles. This led to the innovation of additive manufacturing technologies, which were originally called time compression technologies and rapid prototyping technologies. Even though casting, welding, and powder metallurgy also appear to be additive manufacturing processes, they are not true near-net-shape processes because they are either preceded or followed by additional process steps in realizing the final product. In contrast, the newly invented additive manufacturing process directly converts the computer aided design solid models into almost ready-to-use products or even a product assembly without any limitation on geometric complexity.

The next two sections present the subtractive manufacturing processes and additive manufacturing processes with a particular emphasis on their relative process capabilities and material and energy sustainability potential. The section following them presents the hybrid additive and subtractive processes.

3.2 Subtractive manufacturing processes

In the following subsections, the subtractive manufacturing processes are discussed succinctly. A brief process description is given for each process with an emphasis on material and energy sustainability. A comprehensive treatment of these technologies can be found in several excellent textbooks including one by Kalpakjian and Schmid (2020).

3.2.1 Metal casting

Metal casting is used for manufacturing primary or intermediate stage products that are later processed by other manufacturing processes to produce the final product. Casting is a near-net-shape process because the relative volume of the appendages as compared to the volume of the product is small. However, casting is seldom the final process except in a few specialized processes such as shell casting. Casting involves melting metal or alloy and pouring into a sand mold cavity of the required shape. Since it is often difficult to prepare a mold cavity of a suitable parting plane to the exact geometry of the product, a simplified overall shape is first cast, resulting in the *casting*. The casting is also often left with a few appendages such as sprues, risers, and gating systems that need to be removed and scrapped. This mandatory finishing involved in the casting process produces wastage of material even though it is not strictly a subtractive process. The trimmed casting can be further processed by forming, machining, or welding to obtain the final product. Figure 3.1(a) shows the casting process. Figure 3.1(b) shows the injection molding process, which is one of the most widely used processes for making macro to large size thermoplastic parts.

The casting process can incur significant material wastage in the form of sprue bases, sprues, runners, gates, and risers. The percentage ratio of the weight of the final finished casting to the sum of the casting weight and appendage weight can be defined as the yield, and it is in the range of 86%–88% for typical castings (Sun *et al* 2008). Multi-objective optimum mold cavity design taking into account the yield, in-fill velocity, and shrinkage porosity can result in reduction of wastage by about 11% (Zhang *et al* 2017).

3.2.2 Metal forming

Metal forming involves the application of mechanical force to induce a permanent change of shape through plastic deformation in metals and alloys. Figure 3.1(c) shows the metal extrusion process, which is one of the most widely used hot metal forming processes. Any metal or alloy initially undergoes elastic deformation upon loading, which reversible. Once the elastic limit is crossed, plastic deformation ensues and the force required to cause further plastic deformation increases due to strain hardening, as is evident from the simple tension test force–displacement relation (figure 3.1(d)). The load or stress reaches a maximum point, called the ultimate tensile strength point, up to which the deformation and reduction in cross-sectional area due to Poisson's effect are uniformly distributed over the entire length of the part.

After the ultimate point is reached, the strain become localized to a single weakest cross-section, hence resulting necking and subsequent fracture.

Metal forming is an energy-intensive process and involves bulky processing equipment because the equipment has to withstand large deformation loads with minimum compliance based on the material being processed (Zhao *et al* 2015). Parts made of soft metals such as aluminum, magnesium, tin, copper, and mild steel are

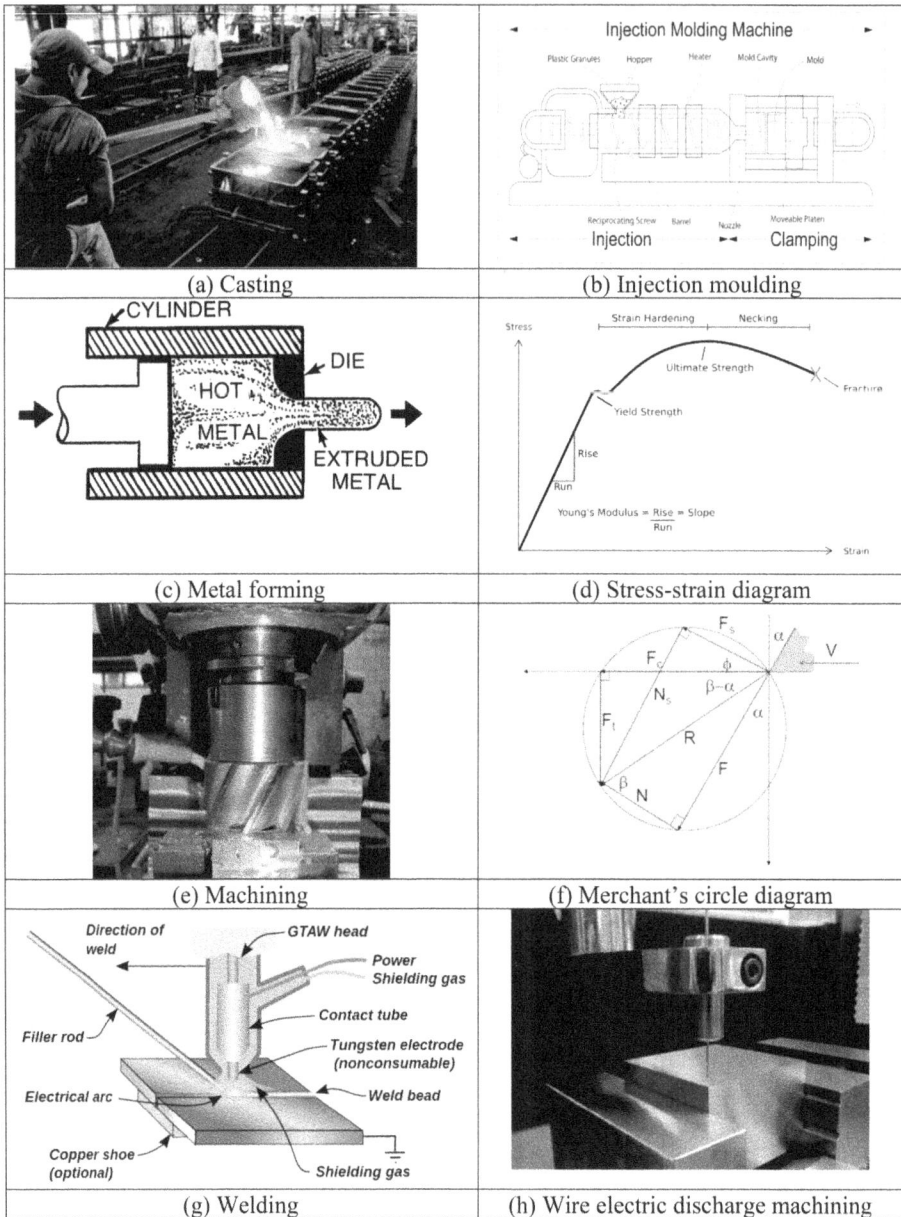

Figure 3.1. Machining and other traditional manufacturing processes. The wire electric-discharge machining (WEDM) process (see (h) above) is a non-conventional process but is still based on the subtractive principle.

relatively easier to process using metal forming processes but high strength metals and alloys such as titanium and Inconel are difficult to process by metal forming.

Metal forming is a necessary secondary processing stage after casting for most products and almost 70% of metal products may have undergone at least one metal forming process to achieve their initial shape (Zhong *et al* 2012). Metal forming is a near-net-shape manufacturing process involving minimal wastage material in the form of flash, which is the extra volume of metal projecting from the punch cavities. Therefore, metal forming cannot be strictly called a subtractive process. Nevertheless, since metal forming can never produce a finished product all by itself but needs to be followed by machining and sometimes welding to produce the final product, it needs to be seen in association with the latter two processes.

3.2.3 Machining

Machining is purely a subtractive process of achieving the shape of the product from the raw material. The process depends primarily on the mechanics of material removal and it is not usual for certain complex parts to require as much as 70% of material to be removed as waste in the form of chips. Therefore, machining has high specific energy consumption and high material wastage because the chips cannot be reused for any purpose but need to be sent for remelting or disposed of as scrap. Figure 3.1(e) shows the face milling process, which is used to accurately obtain a flat surface on a part of medium to large size. Figure 3.1(f) shows the Merchant's circle of machining forces, which plays a critical role in designing the machining process and selecting a suitable machine tool for a given part of specific material and size.

The raw material stock for machining often arrives from the metal forming section in the form of billets of suitable shape, such as block-shaped or cylinder-shaped based on the overall shape of the product to be produced, prismatic or rotational, respectively. The reduction of the raw material into the final product is normally 'process planned' into a set of successive machining processes, each of which produces one feature and may require a separate machine tool. In a typical machine shop, different manual and automated machine tools are arranged in a specific layout and the part is routed through these machine tools in a sequence as per the process plan for imparting the features. The time required for movement of the part between the machine tools is called as the work-in-process (WIP) and is one of the main components of wastage of productivity.

In recent decades, machining technology has seen tremendous technological developments and several new manufacturing processes have been invented to overcome the need for processing of new materials and achieve very small sized holes, complex geometric slots, stricter quality requirements, and very high finish (Kalpakjian and Schmid 2018). Current day subtractive manufacturing processes can be best understood by categorizing them into three groups, namely (i) traditional macro-machining processes (turning, drilling, milling, shaping, slotting, boring, etc), (ii) non-traditional machining processes (abrasive jet machining, water jet machining, electric-discharge machining, abrasive flow machining, chemical machining, lithography, etc), and (iii) micro-machining (micro-drilling, micro-milling, etc).

In the traditional machining processes, the fundamental rule is that the tool is always harder than the workpiece and the tool has to be in physical contact with the workpiece over a well-defined interface geometry to achieve the planned material removal. These constraints were overcome in non-traditional machining (or unconventional machining processes), in which not only a harder tool, but even, for example, water can be converted into a cutting tool for machining a very high strength or highly brittle material by manipulating the energy and transportation of the water. In micro-machining processes, features of very small size (of the order of a few microns) can be machined in a part of overall size of less than 10 mm by accurately predicting the size-effects and designing the process innovatively. Figure 3.1 shows some most popular machining processes.

3.2.4 Welding

Welding is not a complete manufacturing process in itself but it is a fabrication technology to join two metallic parts made by other processes to produce the final product. Having said that, it must be accepted that industrially welding is a crucial manufacturing technology in many sectors including automotive, aerospace, nuclear power, and consumer products. The welding process offers a set of versatile technologies using which two parts made of similar metals (metal arc welding, tungsten arc welding etc, figure 3.1) or dissimilar metals (brazing and soldering) can be joined in a variety of configurations. Welding is a power-intensive process and consumes a lot of electric or gas energy. Welding is a near-net-shape process and involves little or no wastage of material and therefore it is not strictly a subtractive process.

3.3 Additive manufacturing processes

Additive manufacturing (AM) is a disruptive technology invented in the 1980s to face the new challenges of manufacturing newer materials, with minimum lead time requirements, geometric complexity of parts, and mass customization of products. AM was initially available only for polymers for rapid prototyping processes for design verification. The technology grew quickly thanks to advanced computers and computer aided design technologies and soon metal additive manufacturing became possible. Several leading organizations in the world, including GE and Siemens, have accepted the promise of AM in improving the design of gas turbine parts to reduce the number of components.

The unique and fundamental principle of additive manufacturing is to build the *entire part* or assembly layer-by-layer by deposition on a build plate of the material or energy or both. This is unlike welding, where only the joint portion between the two parent metals, which already may be fully finished portions of the final product, is built in single or multi-pass material and energy transfer. Therefore, even though several techniques and computational procedures already developed for welding over many years may be useful in constructing the same for metal additive manufacturing, AM is a distinct process and involves unique challenges of process and design considerations.

AM takes the product data information directly from a computer aided design (CAD) model in a specific format, which may be STL or OBJ or another. These neutral formats are designed to be very simple in syntax and consist of mainly the geometry and only limited topology (for example the STL format) information. The AM machine-specific software implements a mathematical slicing algorithm to convert the solid model of the product in the neutral format into layer by layer slices and, in turn, a numerical control (NC) path planning for the moving elements in the AM machine. The moving elements, based on the type of AM process, may consist of a laser head or electron beam gun or extrusion head for the supply of energy or material or both, and the build table. Most AM processes produce parts requiring post-processing, the extent of which depends on the actual process, for improvement of strength, surface finish, or removal of the support material.

Currently, as per the ISO, today AM technologies can be broadly classified as (i) powder bed fusion (PBF) processes (figure 3.2(a)), (ii) directed energy deposition (DED) processes (figure 3.2(b)), (iii) extrusion type processes (figure 3.2(c)), and (iv) photopolymerization technologies (figure 3.2(d)).

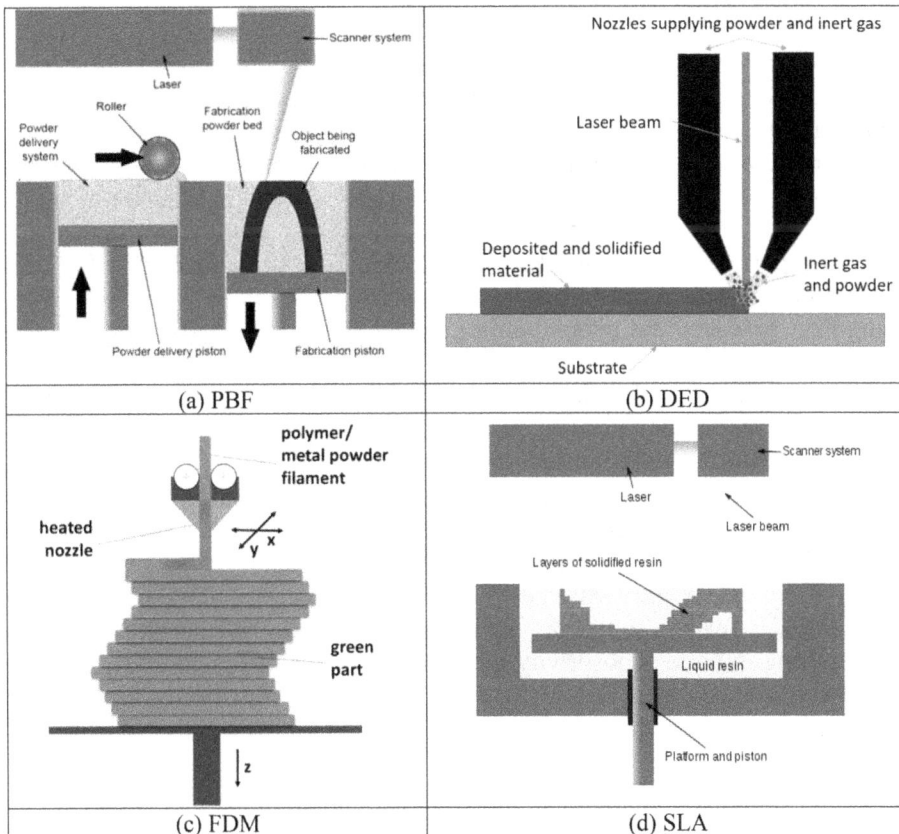

Figure 3.2. The four categorizes of additive manufacturing technologies standardized by the ISO: (a) powder bed fusion (PBF), (b) directed energy deposition (DED), (c) extrusion-based (fused deposition modeling (FDM)), and (d) vat photopolymerization (stereolithography (SLA)).

3.3.1 Powder bed fusion processes

In powder bed fusion (PBF) processes the raw material in the form of a powder of particle sizes ranging from 1 to 50 μm is spread on the build plate layer after layer, and each layer is either sintered or melted and solidified by the heat energy supplied by a laser beam or electron beam.

The powder bed fusion process was first invented for thermoplastic powders as selective laser sintering. Sintering is a process by which powder particles melt only partially due to the heat supplied by the laser on their surfaces, then flow and form necks among the particles at the contact points, thereby producing a porous part with a relative density of about 90%–95%. For increased densities, the parts can be put in an oven to further facilitate consolidation and closure of the internal pores. As metal additive manufacturing grew in relevance for critical strength-related applications such as aerospace engineering, porous parts became less relevant and hence a full melting mechanism in the form of selective laser melting was introduced into AM machines. This technology was eventually standardized as the generic powder bed fusion (PBF) processes, in which the energy to the powder bed may be supplied either by a laser or an electron beam.

A large range of metals and alloys, including AlSi10mg, Ti–6Al–4V, Inconel625/718, and Co-Cr can be currently processed by the existing PBF technologies.

The minimum size of a feature achievable in PBF is a function of the laser beam diameter and the layer thickness. Laser beam diameters as small as 40 μm are now achievable. However, the smallest layer thickness obtainable depends on the powder particle size range (Tan *et al* 2017, Tiberto *et al* 2019). Therefore, there has been a consistent effort in the research to explore the possibilities of making finer powders and understanding the effect of powder size on the process capability.

As the powder particles become finer in size, several enhancements in the process effectiveness and product quality are observed. The amount of energy required to produce a part can be estimated approximately as the product of the amount of heat required by one particle and the number of particles. As the power particles become finer, the energy to melt one particle decreases more dramatically than the effect of the increase in the number of particles on the overall energy required for a given part volume. This improvement in the kinetics of melt consolidation is further supported by an increase in the surface-area-to-volume ratio, resulting in better abosptivity of laser heat energy (Simchi 2004). The heat absorptivity becomes more critical for those metals and alloys with shiny surfaces, such as AlSi10mg, that tend to reflect the laser incidence. Normally Nd:YAG of 10 μm wavelength is used instead of a CO_2 laser of 100 μm to improve the absorptivity. Finer powder also improves the flowability (Spierings *et al* 2016) and packing density (Spierings and Levy 2009, Elliott *et al* 2016) in PBF. The lower limit for powder particle size is imposed by the considerations of low-ignition energy and oxidation.

3.3.2 Directed energy deposition processes

The directed energy deposition (DED) processes differ from PBF in that both the material (either powder or wire) and energy are externally supplied to the build platform.

Over the last few years DED has been developed to be much more versatile with five or seven axis machines and sophisticated technology with sensor-based feedback systems. The DED process permits the supply of either an equi-atomic mixture of elemental powders of constituents or granules of pre-alloyed powder. Providing elemental powders helps achieve flexibility in changing the relative compositions for studies and is less expensive (Preisler *et al* 2023). Greater control of process parameters is essential when elemental powders are used. The scanning velocity of the laser head can have a profound influence on the microstructure and mechanical properties of the part. A lower scanning velocity increases the depth of heat penetration and improves the homogeneity of melting of all alloying elements when elemental powders are used.

The DED is in particular more suitable that PBF for the repair of worn machine components that are too expensive to be scrapped. The virtue of DED in this respect is the five axis capability with which the DED head can be numerically controlled through both translations and rotations to follow the complex geometry of the surface to be repaired. Previously, oxyacetylene or arc welding were options for this type of repair but due to their excessive heat input that resulted in brittle microstructures, DED is being seen as a superior alternative for its better control of heat input and geometric traverse. On the other hand, DED was found to introduce residual stresses in the repaired structures necessitating post-processing heat treatment for acceptable fatigue life (Hamilton *et al* 2023). The effect of laser power intensity and distribution on the microstructure and mechanical properties can be assessed using the thermal gradient (G) at the liquid–solid interface and the solidification rate (R) (Bremer *et al* 2023), the ratio (G/R) of which determines the grain morphology and the product ($G.R$) of which determines the grain size. The parameters G and R can be estimated by the following equations using the temperature distribution measured by *in situ* imaging using thermal cameras:

$$G = |\nabla T|$$

$$R = v\frac{\nabla T}{|\nabla T|},$$

where v is the laser feed rate.

3.3.3 Material extrusion processes

Fused filament extrusion of thermoplastics has been one of the most successful and popular additive manufacturing technologies due to its low machine cost, applicability to a wide range of thermoplastics, and minimal post-processing requirements. The technologies built around this process are popularly known as fused filament fabrication (FFF) and fused deposition modeling (FDM). The structural integrity and mechanical strength of an FDM part depends on the quality of raw material filament used, in terms of its freedom from dissolved gases and air pockets. A poor quality filament releases the dissolved gases during the fusion and the gases and any air pockets become entrapped in the solidified part and manifest themselves as a defects called *lack of fusion* (LOF) defects. More recently, *in situ* imaging techniques are being developed to capture the presence of LOFs and other defects in each layer of the part. The defect images can then be used to train suitable pre-trained artificial neural networks or convolutional neural networks (CNN) in transfer learning mode. A simple yet effective image capturing experimental set-up was developed in a recent publication and corresponding image data of the layers of a tensile testing specimen were used to train the AlexNet CNN and use it to predict three different types of defects, namely the LOF, spikes, and mismatch, with good accuracy levels (Regalla *et al* 2022).

Using similar machine learning (ML) based interventions, more comprehensive structural health monitoring systems for FDM process have been developed recently (Fu *et al* 2023). These monitoring systems can also predict the seriousness of defects in the system as per a criterion and stop the process for process parameter correction only if the defect size is considerable.

FDM is an AM technology in which the polymer parts made using the process find both prototyping applications and critical finished product uses. For example, FDM-made polylactic acid (PLA) thermoplastic acoustic sandwich panels were studies for noise reduction in aerospace applications (Pierre *et al* 2023). Topologically optimized heat sinks made of thermoplastics may not offer a lot of mechanical strength but can offer exceptional thermal properties when they are made by FDM with thermally conducting polymers, such as linear low density polyethylene (LLDPE) doped with a conducting filler such as graphene nano-particles (Huttunen *et al* 2022).

Adapting the extrusion type AM processes to metals and alloys seemed impossible until recently due to the difficulty of finding a machine architecture that can withstand, at low cost, the high melting point temperatures of metals and alloys. However, several innovative methods have been studied in recent years to indirectly use extrusion type processes for metal additive manufacturing. For example, FDM of metal particles held together in the raw material filament by a thermoplastic binder is becoming established as material extrusion additive manu-facturing (MEAM). This process consists of three stages. In the first stage of *shaping*, the metal particle laden thermoplastic filament is extruded through a nozzle and deposited as the required part on the build platform just in the same way as in the traditional FDM process. The deposited part is called a green part because it has poor mechanical strength as the thermoplastic binder and the metal particles in it do not have any direct bonding between them. The green part is then taken to a furnace to carry out the second stage of *debinding*, where the thermoplastic material in the part is almost fully melted out or evaporated out, after which the part is called a brown part. The brown part is finally subjected to a sintering process in a furnace wherein, due to atomic diffusion, the metal particles, now held together in the shape of the desired part, develop necks among their contacts and grow in relative density to as high as 99% (Spiller *et al* 2022).

3.3.4 Photopolymerization processes

The vat polymerization or stereolithography process was among the first additive manufacturing processes to be commercialized successfully and is a photopolymerization process. Liquid resins mixed with catalysts that promote polymerization upon exposure to a particular wavelength of laser light are the raw materials in photopolymerization technologies.

A significant development in vat polymerization technologies in recent times has been four-dimensional (4D) printing. In 4D printing, the liquid resin used is a shape changing polymer (SCP) and after it is imparted with the required pre-designed shape through the regular vat polymerization process, the fourth dimension of shape change can be realized when the part is exposed to an external stimulus such as heat or electromagnetic fields. 4D printing can also be achieved using FDM, but the advantage of vat polymerization is its ability to process liquid polymers that result in parts having elastic flexibilities whereas most thermoplastics used in FDM are rigid in nature.

Other important attributes of vat polymerization made parts have been their high surface finish and fine features due to very small layer thickness, down to 16 μm or less, which is not possible in FDM. These superior features are exploited in medical device manufacturing (Rey-Joly Maura *et al* 2021, Chen *et al* 2019) and point-of-care microfluidic channel devices (Naula *et al* 2021).

3.4 Hybrid additive and subtractive processes

The major advantage of AM over traditional manufacturing technologies, in particular over the subtractive manufacturing processes such as machining, stems from its capability to reduce material waste. The *buy-to-fly* (Kobryn *et al* 2006) ratio, which is defined as the ratio of the volume of the raw material billet to the volume of the finished product, for some the machined aerospace components can be as high as 10:1, indicating the high wastage of material. The selection of manufacturing process for a given part needs to be based on suitable criteria. Several different criteria have been proposed previously in the literature. One of them is based on the amount of energy consumption in the manufacturing stage of the life cycle. The advantage of this criterion is that it includes concerns around sustainability (Watson and Taminger 2018).

Despite the great promise offered by metal AM, the technology is still developing and cannot fully replace the existing subtractive processes for at least another two decades. Even after that, AM will establish itself as a viable process for job-shop or small batch high variety production of high geometric complexity parts whereas the conventional manufacturing techniques will be suitable for mass production. Therefore, the sensible step an industrial engineer can take in deploying these manufacturing technologies is to look for the combined best of additive AM and subtractive manufacturing (SM), an area in which there has been a lot of interest in recent times.

Some of the efforts in developing hybrid AM–SM have focussed on proposing a framework for an integrated formal hybrid process (Manogharan *et al* 2015) whereas others have tried to combine AM and SM with specific features made by each of them as is suitable in particular application categories (Watson and Taminger 2016). In a single dedicated hybrid AM–SM workstation that implemented a selective laser melting type PBF process and a computer numerical control (CNC) milling process, a nickel maraging steel injection molding mold with complex geometry was manufactured in several stages where AM and SM were used alternately in each stage (Du *et al* 2016). A high part density close to 99.2%, a high surface finish, high dimensional tolerances, and better microstructure than casting were achieved in this hybrid AM–SM processing. Such success in hybrid AM–SM is leading to a more confident philosophy of integrated AM–SM workstations (Flynn *et al* 2016).

3.5 Hybrid additive manufacturing of sensors and micro-devices

Current day sensors and actuators are generally understood to be micro-devices, even though traditionally they can be of any size. Thus a sensor/actuator is synonymous with a micro-sensor/micro-actuator if at least one of its dimensions is in the range of a few micrometers. Sensors and actuators are similar in that both of them convert energy of one form to another form. The difference is that in the case of sensors knowing the output information itself is central to the purpose, and in the case of actuators using that information for creating an event is important (Madou 2002). Sensors can be used to process energy sensing in a variety of fields including mechanical, biological, electrical, and chemical.

The manufacturing needs of micro-devices has given rise to the consideration of two different paradigms. One is the top-down approach where a raw material of sufficiently larger size than the final product is subjected to successive subtractive processes to remove material until the final geometry is obtained. Until the recent development of additive manufacturing, micro-device manufacturing was largely through top-down approaches using cutting, etching, and machining type processes. The wet and dry bulk micro-machining, surface micro-machining, LIGA, and electric discharge machining (EDM) are some of the remarkable technologies developed to meet the needs of high precision miniature machine manufacturing in the last 50 years. Much of this development was helped by the already existing techniques in integrated circuit electronics manufacturing. The bottom-up techniques were limited to a self-assembly monolayer (SAM) method of building parts, which was unfortunately not sufficiently scalable in terms of meaningful time and energy consumption.

Additive manufacturing is a bottom-up method and has great scope for further developments in micro-device manufacturing. This is in particular true with the continuously decreasing layer thicknesses possible in metal AM. Extrusion type AM processes are more difficult to use to make micro-devices due to their limiting minimum layer thickness, which cannot be less than about 200 μm. However, the SLA process for polymers and the PBF and DED processes for metals and alloys can achieve layer thicknesses as low as 10 μm and hence at least few layers can be incorporated into every feature of the micro-device with sufficient strength and stiffness.

One class of micro-devices that has been recently and successfully made using AM are biosensors. Silver nanowire (AgNW) embedded x-pentadecalactone-co-e-decalactone (PDL) matrix nanocomposite strain-gauge biosensors were manufactured by a pneumatic pressure assisted extrusion type AM process (Britton *et al* 2021). These biocompatible sensors can be printed on flexible substrates and can be integrated into wearable or implantable devices for monitoring the health of humans.

Sensors often consist of a non-conducting, geometrically complex, and customized base of polymer on top of which a thin transduction layer made of piezoelectric material and wire connections are deposited. To meet this need, an extrusion type additive manufacturing process is used to make the base. The sensing layer can be made by a variety of 3D printing methods engineered in recent years, including ink-jet printing (IJP), direct ink writing (DIW), and aerosol jet printing (AJP). The minimum layer thickness limitations of 3D printing can also be overcome by resorting to such hybrid processes wherein a secondary process is used to achieve the higher resolution features on the surface of the micro-device. Micro-electro-mechanical system (MEMS) accelerometers were made using such a hybrid AM and 3D printing process recently (Bernasconi *et al* 2022, Fritz *et al* 2022). The SLA can achieve excellent geometric details on additively manufactured solid parts with photo-curable liquid resin as the raw material, which can be used to make the base in the hybrid additive process.

The raw materials for 3D printing of sensors and other micro-devices require special properties, mainly due to the requirements of their applications. For example, biosensors require biocompatibility and cytocompatibility of the material, and sensors to be used on load-bearing machine elements require mechanical strength. Thermoplastic plastics (used in extrusion type 3D printing) and thermosetting liquid resins (used in photopolymerization based stereolithography) alone cannot provide these properties. Nanofiber infused/reinforced nanocomposites have been proposed as solution to overcome this need. Silver nanowire (AgNW) reinforced PDL (Britton *et al* 2021), carbon nanofiber reinforced polylactic acid (PLA) (Hernandez *et al* 2022), and silver nanoparticle reinforced polymer ink for aluminum alloy substrates for strain-gauges for nuclear fuel cladding applications (Phero *et al* 2022) are some examples for these special raw materials. Neodymium (NdFeB) magnetic particle reinforced thermoplastic polyurethane (TPU) filaments for 3D printed sensors for monitoring bone rehabilitation have been fabricated (Shi *et al* 2022).

3.6 Summary

In this chapter subtractive manufacturing technologies and additive manufacturing technologies and their relative process capabilities have been discussed, culminating the newly emerging philosophy of hybrid AM–SM processes. The focus was in particular on saving material and energy in the processes. Based on the existing trends in the AM and SM technologies, it is predicted that the next two decades will see stronger integration between them to realize the best quality complex geometry parts rapidly with both customization and good production rates becoming achievable.

References

Bernasconi R, Hatami D, Hosseinabadi H N, Zega V, Corigliano A, Suriano R, Levi M, Langfelder G and Magagnin L 2022 Hybrid additive manufacturing of a piezopolymer-based inertial sensor *Addit. Manuf.* **59** 103091

Bremer S J L, Luckabauer M and Römer G R B E 2023 Laser intensity profile as a means to steer microstructure of deposited tracks in directed energy deposition *Mater. Des.* **227** 111725

Britton J, Krukiewicz K, Chandran M, Fernandez J, Poudel A, Sarasua J-R, FitzGerald U and Biggs M J P 2021 A flexible strain-responsive sensor fabricated from a biocompatible electronic ink via an additive-manufacturing process *Mater. Des.* **206** 109700

Chen L, Lin W-S, Polido W D, Eckert G J and Morton D 2019 Accuracy, reproducibility, and dimensional stability of additively manufactured surgical templates *J. Prosthet. Dent.* **122** 309–14

Du W, Bai Q and Zhang B 2016 A novel method for additive/subtractive hybrid manufacturing of metallic parts *Procedia Manuf.* **5** 1018–30

Elliott A M, Nandwana P, Siddel D H and Compton B 2016 A method for measuring powder bed density in binder jet additive manufacturing process and the powder feedstock characteristics influencing the powder bed density *Report* Oak Ridge National Lab (ORNL), Oak Ridge, TN

Flynn J M, Shokrani A, Newman S T and Dhokia V 2016 Hybrid additive and subtractive machine tools—research and industrial developments *Int. J. Mach. Tools Manuf.* **101** 79–101

Fritz G, Alshawabkeh M and Faller L-M 2022 Additively manufactured soft linear sensor *Mater. Today Proc.* **70** 201–5

Fu Y, Downey A R J, Yuan L and Huang H-T 2023 Real-time structural validation for material extrusion additive manufacturing *Addit. Manuf.* **65** 103409

Hamilton J D, Sorondo S, Li B, Qin H and Rivero I V 2023 Mechanical behavior of bimetallic stainless steel and gray cast iron repairs via directed energy deposition additive manufacturing *J. Manuf. Processes* **85** 1197–207

Hernandez J A, Maynard C, Gonzalez D, Viz M, O'Brien C, Garcia J, Newell B and Tallman T N 2022 The development and characterization of carbon nanofiber/polylactic acid filament for additively manufactured piezoresistive sensors *Addit. Manuf.* **58** 102948

Huttunen E, Nykänen M T and Alexandersen J 2022 Material extrusion additive manufacturing and experimental testing of topology-optimised passive heat sinks using a thermally-conductive plastic filament *Addit. Manuf.* A **59** 103123

Kalpakjian S and Schmid R 2018 *Manufacturing Engineering and Technology* (New York: Pearson)

Kalpakjian S and Schmid S R 2020 *Manufacturing Engineering and Technology* 8th edn (New York: Pearson)

Kobryn P A, Ontko N R, Perkins L P and Tiley J S 2006 Additive manufacturing of aerospace alloys for aircraft structures *Report* Materials and Manufacturing Directorate, Air Force Research Lab, Wright-Patterson Airforce Base, OH

Madou M J 2002 *Fundamentals of Microfabrication: The Science of Miniaturization* (New Delhi: CRC Press)

Manogharan G, Wysk R, Harrysson O and Aman R 2015 AIMS—a metal additive-hybrid manufacturing system: system architecture and attributes *Procedia Manuf.* **1** 273–86

Naula E A, Rodríguez B L, Garza-Castañon L E and Martínez-López J I 2021 Manufacturing of stereolithography enabled soft tools for point of care micromixing and sensing chambers for underwater vehicles *Procedia Manuf.* **53** 443–9

Phero T L, Novich K A, Johnson B C, McMurtrey M D, Estrada D and Jaques B J 2022 Additively manufactured strain sensors for in-pile applications *Sens. Actuators* A **344** 113691

Pierre J, Iervolino F, Farahani R D, Piccirelli N, Lévesque M and Therriault D 2023 Material extrusion additive manufacturing of multifunctional sandwich panels with load-bearing and acoustic capabilities for aerospace applications *Addit. Manuf.* **61**

Preisler D, Krajňák T, Janeček M, Kozlík J, Stráský J, Brázda M and Džugan J 2023 Directed energy deposition of bulk Nb-Ta-Ti-Zr refractory complex concentrated alloy *Mater. Lett.* **337** 133980

Regalla S P, Kaushal A and Khetan S 2022 Machine learning (ML) based prediction of defects in extrusion type additively manufactured parts *3rd International Conference on Future Technologies in Manufacturing, Automation, Design and Energy (NIT, Puducherry, India, 13–16 Dec)*

Rey-Joly Maura C, Godinho J, Amorim M, Pinto R, Marques D and Jardim L 2021 Precision and trueness of maxillary crowded models produced by 2 vat photopolymerization 3-dimensional printing techniques *Am. J. Orthod. Dentofacial Orthop.* **160** 124–31

Shi Y *et al* 2022 Additive manufactured self-powered mechanoelectric sensor as the artificial nucleus pulposus for monitoring tissue rehabilitation after discectomy *Nano Energy* **96** 107113

Simchi A 2004 The role of particle size on the laser sintering of iron powder *Metall. Mater. Trans.* B **35** 937–48

Spierings A B and Levy G 2009 Comparison of density of stainless steel 316L parts produced with selective laser melting using different powder grades *Proceedings of the Annual International Solid Freeform Fabrication Symposium (Austin, TX)* pp 342–53

Spierings A B, Voegtlin M, Bauer T and Wegener K 2016 Powder flowability characterisation methodology for powder-bed-based metal additive manufacturing *Prog. Addit. Manuf.* **1** 9–20

Spiller S, Kolstad S O and Razavi S M J 2022 Fabrication and characterization of 316L stainless steel components printed with material extrusion additive manufacturing *Procedia Struct. Integr.* **42** 1239–48

Sun Z, Hu H and Chen X 2008 Numerical optimization of gating system parameters for a magnesium alloy casting with multiple performance characteristics *J. Mater. Process. Technol.* **199** 256–64

Tan J H, Wong W L E and Dalgarno K W 2017 An overview of powder granulometry on feedstock and part performance in the selective laser melting process *Addit. Manuf.* **18** 228–55

Tiberto D, Klotz U E, Held F and Wolf G 2019 Additive manufacturing of copper alloys: influence of process parameters and alloying elements *Mater. Sci. Technol.* **35** 969–77

Watson J K and Taminger K M B 2018 A decision-support model for selecting additive manufacturing versus subtractive manufacturing based on energy consumption *J. Clean. Prod.* **176** 1316–22

Zhang L, Belblidia F, Davies H M, Lavery N P, Brown S G R and Davies D 2017 Optimizing gate location to reduce metal wastage: Co–Cr–W alloy filling simulation *J. Mater. Process. Technol.* **240** 249–54

Zhao K, Liu Z, Yu S, Li X, Huang H and Li B 2015 Analytical energy dissipation in large and medium-sized hydraulic press *J. Clean. Prod.* **103** 908–15

Zhong J, Zhao K, Liu Z F and Li X Y 2012 Review of the research on low carbon manufacture of metal-forming equipment and future development *J. Hefei Univ. Technol. (Nat. Sci.)* **35** 1594–600

IOP Publishing

3D Printed Smart Sensors and Energy Harvesting Devices
Concepts, fabrication and applications
Sanket Goel and Sohan Dudala

Chapter 4

3D printing materials for sensors and their commercial availability

Prakash Katakam, Madhavi Lakshmi Ratna Bhavaraju and Shanta Kumari Adiki

4.1 Introduction

Sensors are technological devices that are employed in many walks of life. For these products, one design parameter, their size, is crucial for their functioning, efficiency, and suitability to be incorporated into functional components or parts [1]. A sensor, in the simplest terms, is a device that converts physical inputs such as temperature, pressure, humidity, pH, magnetic field, and so on, into an electrical signal. Sensors are used in biomedical and health care applications such as the measurement of heart rate, blood pressure, oxygen levels in the blood, nerve impulses, pulse rates, and body temperature, and in food safety and quality testing [2]. Engineering and technological applications include the detection of speed, smoke, metals, chemicals, and heat, automation applications, earthquake sensors, proximity sensors, and so on [3].

The emergence of 3D printing (3DP) technology offers the benefits of producing products with complex designs in small sizes, with great precision and accuracy. The versatility of 3DP is reflected in the variety of manufacturing approaches available. Vat photopolymerization (VP), material jetting (MJ), material extrusion, directed energy deposition (DED), sheet lamination (SL), and powder bed fusion (PBF) are some of the techniques that can be used for 3DP or additive manufacturing (AM) [4]. The most advanced 3DP technologies, transduction sensing systems, and 3DP for food quality and safety were reviewed recently [5].

3DP of sensors is an important area of study [6]. Sensor components at a basic level include (i) an input/receiver or a sensing unit, (ii) an integrated circuit or a processing unit, and (iii) an output or display or the communication unit. Wireless sensors have the additional component of a power source [7]. The housing is the part of the sensor that encases its components. The 3DP of a sensor may be directed towards making the housing, the sensing element, or any other part. For 3DP of sensors, suitable materials are chosen keeping in mind their functionality and the

doi:10.1088/978-0-7503-5351-9ch4

environment in which they will be used. Materials may be broadly classified into plastics, polymers, metals, ceramics, and miscellaneous [8].

Polymers such as polydimethylsiloxane (PDMS) [9], polyethylene terephthalate (PET) [10], and polyimide (PI) [11], have been employed for the manufacture of substrate components in flexible sensor prototypes, while the electrode portion includes carbon nanotubes (CNT) [12], graphene [13], and gold nanoparticles [14]. The mechanical, electrical, and thermal properties of the prototypes are ultimately determined by the unification of these distinct polymeric substrates [15] and electrodes [16].

The introduction of flexible sensors has resulted in several modifications to the characteristics of the prototypes, thus mitigating some of the disadvantages of sensors based on silicon. Among the numerous manufacturing processes for generating flexible sensors, photolithography [17], inkjet printing (IJP) [18], screen printing [19], laser cutting [20], contact printing [21], and 3D printing [22, 23] are often utilized. This chapter gives a brief overview of sensors. It mainly focuses on the materials used for 3DP of sensors and provides information on their commercial availability.

4.2 Types of sensors

Sensors may be classified as given below based on the input to which they respond. Their functionality suggests the material that would be used in their 3DP.

4.2.1 Strain sensors

Strain sensors or strain gauges that convert compressive or tensile stresses into electrical signals have been developed and used in a variety of scientific activities [24]. These sensors are composed of a flexible conduction material that has been implanted in or attached to a stretchy material. Customized conformal and extendable 3D-printed strain sensors have been created using soft matrices [25–28]. Inks for 3DP strain sensors have been made with a variety of basic components, such as nanotubes, graphene, and nanoparticles [29]. Flexible electronics have limitless potential to become one of the major trends in the exploitation of wearable gadgets and electronic skin due to their groundbreaking and pioneering improvements. In flexible, integrated electronic systems, strain sensors, which have received a great deal of attention and serve as significant intermediaries for the gathering of external mechanical data, are considered vital components [30]. The 3DP procedure of an accelerometer-capable construction has been described previously [31]. Briefly, cantilevers were manufactured by printing on PET, resulting in increased flexibility in the manufactured sensors. Silver paste and polyvinyl alcohol (PVA) have been employed as a structural material and a sacrificial substrate, respectively. Miniature sensors to measure micro displacement have been manufactured via 3DP technology [32].

4.2.2 Force sensors

Force sensors have been applied in a vast array of applications. They have been utilized in manufacturing processes, medical devices, logistics, haptic interfaces, the automobile industry, and robotics, among other applications. Using force sensors, applied forces are transformed into electrical signals. This class of sensors typically consists of three major components: a transducer, a flexure, and a packing unit. All three of these primary components may be manufactured using 3DP technology [1].

4.2.3 Pressure sensor

Various pressure sensors can detect and monitor pressure variations, structural stresses, and a goal pressure. Changing the geometric dimensions of a polylactic acid (PLA) printed component made it feasible to measure various pressures. PDMS can be employed to provide device flexibility [33].

4.2.4 Temperature sensors

Temperature sensors have become indispensable in a variety of applications. Similar to other sensor types, temperature sensors may be produced via 3DP. The IJP method of 3DP was utilized to create a flexible temperature sensor in [34].

4.2.5 Particle sensors

Particle sensors, in addition to other types of sensors, have found a number of crucial applications [35]. These sensors can sense particles in the air and can be used to measure pollution. However, they are sometimes inaccurate and susceptible to environmental variables. To detect particles, engineering structures comprising channels and sensors may be constructed utilizing the benefits of 3DP technology.

4.2.6 Tactile sensors

A tactile sensor can be used to detect, transmit, or receive information. This sort of sensor may offer information on contact pressure and is employed in a variety of applications. Since the production method and materials used in touch sensors are costly, they are not suitable for laboratory-specific experiments. Utilizing AM in the production of tactile sensors has increased their adaptability. Soft robotics is gaining popularity due to the rising need for delicate part manipulation on manufacturing lines. Due to its capacity to produce designs with complicated geometries and multimaterial printing capability, 3DP technology is a crucial production tool for soft robots [36].

4.2.7 Miscellaneous

Some examples of sensors are lidar sensors, encoder sensors, motion sensors, biomedical sensors, color sensors, IR distance sensors, ultrasonic sensors, proximity sensors, flow sensors, pH sensors, temperature sensors, current sensors, and touch sensors [37]. These sensors have several industrial applications. Encoder sensors are utilized [38] in printers, robotics, food processing, drilling machines, material handling, medical scanners, dispensing pumps, axis controls, military-grade antennas, and telescopes, for example.

4.3 Examples of AM sensors

Here we look at some potential applications of AM sensors in electromechanical and biomechanical systems.

4.3.1 Piezoresistive method

Using piezoresistive materials, 3D-printed traxel-based strain sensors are suited for use in tactile and force sensing applications due to their high strain sensitivity [39]. These sensors were made using co3DP with conductive pastes, just like the previously reported flexible sensors that used silver palladium paste as the conductive material [40]. Using a blend of silicone elastomer containing silver particles, researchers produced a tiny touch sensor that can detect the pulse of a person using a custom made paste [41]. Strain sensors have further practical uses in sensing devices that employ conductive polymer composites as touch and tactile sensors, produced with soft conductive and nonconductive thermoplastic polyurethane (TPU) materials [42]. The remarkable flexibility of these TPU materials is demonstrated by their use in a glove for monitoring finger flexion [43].

Due to their potential capabilities in biomonitoring applications, 3D-printed pliable composite piezoresistive sensors have contributed to the development of tailored therapeutics. A silicone rubber (SR) matrix is a key choice for 3D-printed products that require flexibility. MJ of high-viscosity conductive materials using a UV cross-linkable SR and powder carbon fibers is well demonstrated [44].

4.3.2 Capacitive method

The capacitive sensing method has an advantage over piezoresistive sensing methods in that the measurement is not affected by drift, nonlinearity, or hysteresis in the conducting parts. The capability to print soft materials enables the production of force sensors consisting of a parallel plate capacitor with a directly compressed dielectric [45]. Note that the capacitive sensors frequently require shielding or protection to decrease the effect of parasitic capacitances. Printing a conductive assembly that generates a capacitive assembly that can sense the variation of the dielectric constant in surrounding materials is another prevalent approach. For example, the capacitance is altered by glass, finger, wood, or steel [46].

4.3.3 Inductive method

In general, the poor conductivity of carbon filaments precludes their use as inductors [47]. Inductors may be 3D-printed via direct printing with an Electrifi filament or electroplating onto Electrifi material [48]. Further, inductive sensors are created by embedding wires of nickel and copper [49], 3DP with silver paste [50] or metal liquid [51], and spraying silver nano-ink [52] onto a graphene filament coil.

Due to the high piezoresistive constant of many 3DP conductors, piezoresistive sensors are frequently the simplest to construct. However, the vast majority of piezoresistive materials exhibit significant nonlinearity; hence, a capacitive sensor could be beneficial in applications where strong linearity is essential, correction is not possible, and opportunistic capacitances may be shielded. When the printing technique permits the production of highly conductive conductors, inductive sensing methods can be added to the list of 3D printable sensors.

4.4 Materials used for sensors

The extrusion technologies mentioned in section 4.1 can be used to print a wide range of components, from extremely elastic to extremely rigid [53, 54]. Various polymers are employed in the manufacturing of sensors. PLA continues to be the most popular and ecofriendly base polymer for the creation of electrochemical components due to its flexibility during printing and reduced printing temperature requirements in comparison to other plastics. As a downside, its mechanical qualities are weak [55, 56]. Acrylonitrile butadiene styrene (ABS) is also often used; however, when printing, it emits poisonous styrene vapors with a disagreeable odor, making its use in a research setting challenging.

PET is an intriguing material that could be used in 3DP because it has greater mechanical resistance than ABS. Thermoplastics such as polytetrafluoroethylene (PTFE) and polyether ether ketone (PEEK) are frequently used in scientific research; however, these materials have melting points above 350°C, and only a small number of expensive 3D printers can handle these polymers [57].

In addition to the above listed polymers, other plastic materials such as nylon, PVA, TPU, thermoplastic elastomer (TPE), polycaprolactone (PCL), polypropylene (PP), polyetherimide (PEI), and Tritan are also employed for 3DP [58]. Although nylon possesses good thermal and mechanical resistance as well as outstanding flexibility, this material readily absorbs moisture and is difficult to print, necessitating high temperatures for extrusion. With comparable extrusion temperatures, TPU/TPE and PP exhibit excellent flexibility and elasticity [3]. However, PP does not readily attach to glass surfaces; therefore certain deformations in the produced material can be visible [59]. PVA, a thermoplastic material, is also a viable choice. This biodegradable polymer is frequently employed as a drug delivery polymer for 3D-printed capsules [60]. PEI, in comparison, requires extremely high temperatures for extrusion [61]. Tritan is an excellent choice where mechanical strength is necessary. Manufacturers believe this material to be the most durable and robust filament on the market for 3DP, but it requires extremely high extrusion temperatures and it is expensive [62]. Additionally, a mix of thermoplastic materials can be used.

Filaments containing composite functional materials have gained popularity. Conductive filaments containing graphene, carbon black, copper, zinc, and steel have become widespread [63]. A conductive substance is required for the preparation of electrochemical sensors and biosensors, making commercially available materials an intriguing choice. However, the high cost, lack of knowledge on their precise composition, and existence of impurities might compromise the reliability and use of some commercial filaments [64].

Alternatively, a number of studies have reported the production of carbon-based conductive filaments that are less expensive than commercial filaments [65].

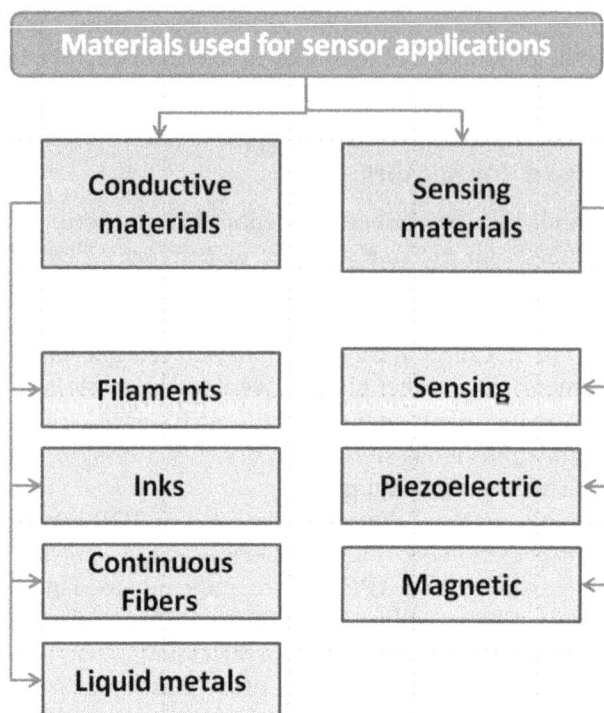

Figure 4.1. Materials used for sensor applications.

The extrusion method can be employed to make such filaments Thermoplastic polymers, such as PLA, ABS, polystyrene (PS), PP, polyethylene terephthalate glycol-modified (PETG), and nylon, and conductive materials, such as graphite, carbon black, grapheme, carbon nanofibers and nanotubes, and several metals, are used to manufacture functional filaments. The carrier polymers and the conductive materials must be mixed thoroughly without the presence of any clumps. The blend can then be heated and passed through the extruder to yield conductive filaments. The proportion of conductive material has to be optimum. The composition of the filament has to be evaluated to preserve printing quality, electrical conductivity, and mechanical and thermoplastic features. Unlike commercial filaments, the manufacture of new filaments facilitates the provision of customized conductivity by varying the quantity of conductive components and by the incorporation of electrocatalysts within the filament to improve electrochemical performance. The materials used for sensor applications in 3DP may be classified broadly as given in figure 4.1.

4.4.1 Conductive materials

The ability to print conductors is crucial for manufacturing sensors-in-a-package components. The conductors can be used to make piezoresistive or capacitive sensors, and they can also be used to wire the sensors to the readout electronics, all inside the same printed structure.

4.4.1.1 Filaments

Filaments contain conducting materials such as graphene, carbon black, CNTs, graphite, or copper powder in a polymer matrix [66]. They are generally printed using the fused deposition modeling (FDM) technique. The conducting materials, also called fillers, are readily available, inexpensive, and chemically stable, making them ideal for filament production [67, 68]. High-filler materials are weaker because of poor particle–matrix adhesion [69]. Moreover, these materials achieve conduction through percolation networks; hence, the influence of filler concentration on resistivity decreases as its concentration increases [70]. Since the materials are composites, they frequently show the piezoresistive effect.

4.4.1.2 Inks

Inks are generally formulated for use in IJP or SP. They are a mixture of conducting materials suspended in a polymeric solution using solvents that dry rapidly. Inks for jetting invariably have a low viscosity (1–30 mPa s), but inks for SP have a significantly greater viscosity (1000–10 000 mPa s) due to their higher solid content [71]. The wetting and viscosity characteristics of the ink dictate how it will interact with other materials after printing. The majority of conductive inks employ silver particles. In applications requiring biocompatibility, poly(3,4-ethylenedioxythiophene):polystyrenesulfonate (PEDOT:PSS) based inks are frequently employed, which are available with varying degrees of flexibility [72]. Some inks require ultraviolet or thermal curing. Having ink that does not dry out or split in the extruder between prints is essential.

4.4.1.3 Continuous fibers

Another class of materials employed for 3DP includes the continuous fiber reinforced composites [73]. To improve the mechanical properties a combination of materials is employed. Filaments are incorporated with filler particles in powder form or short fibers to obtain variable mechanical/conducting properties. However, their mechanical properties are inferior. The incorporation of continuous fibers in the structure of the base material such as nylon or other plastics confers much greater strength to the 3DP structures. Fiber materials for reinforcement include carbon fiber, fiberglass, and copper [74]. Continuous fiber reinforcement in the polymer matrix may be achieved by various techniques such as prepreg, towpreg, and *in situ* polymerization [75].

Due to copper's high wire conductivity, which is equivalent to its bulk conductivity, copper wires may be used to transport huge currents [76]. Carbon fiber is commonly utilized due to its high resistance and yield strength; however, copper wires can be used when high currents or low resistance are required. It is possible to print dry carbon fibers. But as dried carbon fibers are extremely brittle and difficult to work with, they are more commonly used in the form of a prepreg, which is a bundle of polymer-impregnated dry fibers assembled into a filament [77]. Dry fibers with co-extrusion, on the other hand, provide greater freedom in selecting fiber and matrix blends.

4.4.1.4 Liquid metals

Among the conducting materials for 3DP, liquid metals are quite suitable, in particular for incorporation into stretchable sensors. Gallium and gallium alloys such as eutectic gallium indium (EGaIn) and Galinstan (an alloy of gallium, indium, and tin) are an additional class of conductors that can be used to enhance the functionality of sensor materials. These liquid metals can be incorporated into printed products either through direct ink writing or later filling. Due to the metal's liquid state, it can be expanded up to 200% without breaking. It should be noted that liquid EGaIn can tarnish other metals, perhaps causing damage to the printers that use them [78].

From among the types of conducting materials discussed above, the choice of the most appropriate conductive material depends on the application. FDM filaments provide a low-cost alternative for producing actual conductors. In rigid constructions, copper fibers, liquid metals, and silver paints permit a significantly greater current density. In addition to their excellent conductivity, liquid metals and stretchy silver inks may be stretched to a significant extent.

4.4.2 Sensing materials

This section will provide a brief introduction to printable functional materials that may be utilized in the creation of electromechanical and biomechanical sensors.

4.4.2.1 Piezoresistive materials

Piezoresistive devices convert mechanical energy to resistance values. Due to changes in length, diameter, and specific resistivity, the resistance of a stretched conductive material varies. However, in the majority of conductive polymer composites, the resistance, which changes more than predicted by the geometrical effect, results in greater strain sensitivity. This piezoresistive phenomenon is observed in conductive polymer composites due to the transformation of conductive channels in the polymer matrix under strain [79]. A significant disadvantage of conductive polymer composites is the material's nonlinear resistance, which changes under tension. Due to the proper alignment of the conductive elements in the matrix, resistance typically decreases at small stresses and then increases at greater strains [58]. In addition, when the material is being stretched, polymer mobility and failure of the conducting networks occur. At low strain rates, both effects are in equilibrium, leading to negligible changes in resistivity; however, at high strain rates, the failure effect predominates. Additionally, some of the employed polymers, such as TPU, are viscoelastic, which enhances the hysteresis and nonlinearity of the piezoresistor's response [80]. As detailed earlier, it may be possible to simulate these effects and correct for them in software. In addition to conductive polymer composites, piezoresistive sensors might be manufactured with liquid metals, conductive inks/fibers [81, 82].

4.4.2.2 Piezoelectric materials

In contrast to piezoresistive materials, mechanical energy can be transformed into electrical energy and vice versa using piezoelectric components. Typical piezoelectric materials for 3D printing are poly(vinylidene fluoride) (PVDF) and PVDF–ceramic composites [83, 84]. By incorporating copolymer or ceramic particles, the substance may also be dissolved and used as ink [85]. In order to boost the piezoelectric constant, the material is often polarized after production [86]. The piezoelectric crystalline structure of PVDF can be created through printing with a high voltage between both the nozzle and the bed. It should be noted that an electret, which may also be 3DP, can accomplish a comparable effect to that of a piezoelectric material [87].

4.4.2.3 Magnetic materials

There are both soft magnetic thermoplastics and hard magnetic thermoplastics on the market [88, 89]. These thermoplastics may be 3DP and then polarized to create a material with residual magnetism [90]. However, the use of injection-moldable thermo-plastics as opposed to FDM could result in filament which is brittle and hard to control [91]. Definite magnetic thermoplastic composites can be manufactured for magnet and sensor research applications using use wires made of ferromagnetic elements [92].

Piezoresistive materials provide a straightforward and inexpensive method for incorporating pressure sensitivity into 3DP components. In contrast, 3DP piezor-esistors constructed from conductive polymer composites frequently exhibit hyste-resis and high nonlinearities. It is possible to print piezoelectric and magnetic materials. For the majority of applications, however, they must be filed after or during printing.

4.5 AM techniques used for sensor manufacturing

The following AM techniques are used to manufacture sensors [93].

4.5.1 Fused deposition modeling (FDM)

FDM is the most popular among the AM techniques. Printing materials such as PLA, ABS, wax mixes, and nylon are commonly used in FDM. If the printing material is not calibrated appropriately, the manufactured product will lack form and integrity and may leak. Frequently, surface finishing is necessary to obtain the final result. Manufacturing sensor-integrated components relies on the capacity to 3D print conductors. The conductors can be used to make piezoresistive or capacitive sensors and can also be used to wire the sensors to the readout electronics within the printed structure.

4.5.2 Stereolithography (SLA)

In SLA photopolymers are transformed into polymer chains and bonded to the subsequent layer via radicalization. After the procedure is complete, the unreacted resin that serves to hold the structure in place is removed. This SLA method has a resolution as exact as 10 μm, allowing for the printing of high-quality, accurate sensors. One of the best things about this method is that it can be used to make structures with very large volumes.

4.5.3 Polyjet (PJ)

In the PJ technique, a photopolymer is employed to create a three-dimensional model through a photocuring or hardening procedure. In contrast to the FDM technique, which uses a single nozzle for printing, the PJ process employs many nozzles. The print head traverses the x–y-axis of the platform and ejects small droplets of photopolymer to deposit the printing material in a stereolithography (STL) compliant design. One of the primary benefits of this method is that prototypes with a resolution of 16 μm and an accuracy of less than 0.1 mm may be created.

4.5.4 Inkjet printing (IJP)

The 3D IJP involves deposition of powder particles across a platform, via lower viscosity photo-curable resin or hydrogel drops that serve as the printing medium. This liquid medium bonds the powder particles to produce a sufficiently strong solid structure. Each layer may be constructed by ejecting ink through a tiny deposition nozzle, and the 3D object can be constructed layer by layer. One type of ink used in IJP is liquid based, while the other type is wax-based. Adhesion between the binder and powder particles is crucial to preserve the quality of this layered production method [94]. The IJP technology produces the most flexible 3D prototypes [49]. Using this procedure various materials, such as metals, ceramics, polymers, and even live cells, can be deposited onto the prototypes' substrates [95]. Due to its simple working concept and rapid creation of complicated 3D models, this AM approach has a wide range of applications, particularly in the engineering, biomedical, and pharmaceutical fields. Physiological parameters such as body temperature, brain activity, body mobility, blood pressure, heart rate, respiration rate, and skin temperature have been monitored using 3D-printed sensor components incorporated into biomedical equipment [96].

4.5.5 Digital light processing (DLP)

DLP is comparable to SLA, with the exception of the photocuring process. Here, a digital projector screen flashes to show an image-like layer composed of squared voxels. Each 2D hardened layer is generated by exposing liquid polymer to projector light under relatively safe circumstances, as opposed to using several laser scan lines to create a single layer. This procedure is continued until the entire structure has been developed.

4.5.6 Selective laser sintering (SLS)

SLS is an AM technique in which objects are constructed layer by layer. Powder particles are deposited which are then bound together by subjecting them to a laser. Upon this layer, the next layer of powder is deposited and bound. Among the various AM techniques, PJ and SLS procedures are typically utilized for the fabrication of cell culture related sensors. Even though all of these processes are effective to some extent, only IJP and DLP are capable of producing the most repeatable prototypes. PJ technology is now widely used in 4D printing to create prototypes with active components [97].

Table 4.1 gives an overview of AM processes, including the concept, materials, resolution, and biomedical applications of 3DP sensors, and table 4.2 details the

Table 4.1. Overview of the processes of additive manufacturing of 3D-printed sensors. (Adapted with permission from [98]. Copyright 2019 the authors.)

AM method	Principle	Materials	Resolution (μm)			Biomedical application
FDM	Filament extrusion	PLA, ABS, nylon, wax mixtures	x: 100	y: 100	z: 250	Bacteria sensor, cell toxicity sensor, L-DNA sensor, glucose sensor, immunosensor, lactate sensor
SLA	UV assisted polymerization	Acrylate/epoxy resins with photo initiators	x: 10	y: 10	z: 15	Bacteria sensor, cellular sensor, DNA imaging sensor
PJ	Deposition of photo-curable droplets and curing	Polymer	x: 30	y: 30	z: 20	Cell based sensor (ATP sensing), cell imaging sensor, immunosensor, physiological sensor
SLS	Laser-assisted powder sintering	Metallic powder, polyamide, polyvinyl chloride	x: 50	y: 50	z: 200	Cell density sensor
IJP	Ink extrusion on to powder layer leading to powder binding	Photo-resin or hydrogel	x: 10	y: 10	z: 50	Bionic ear, multifunctional bio-membrane
DLP	Curing by a DLP projection onto resin layers by squared voxels	Photo-resins and photopolymers	x: 25	y: 25	z: 20	Glucose sensor, motion control sensor, soft sensor, piezoelectric acoustic sensor

Adapted from Han T, Kundu S, Nag A, Xu Y 2019 3D-printed sensors for biomedical applications: a review *Sensors* **19** 1706 http://dx.doi.org/10.3390/s19071706.

Table 4.2. Details of 3DP sensors: the type of sensor, AM process used, materials used, transduction mechanism, and sensitivity. (Adapted with permission from [6]. Copyright 2022 Elsevier.)

Type of sensor	AM process	Material	Transduction mechanism	Sensitivity
Strain	SLA	Silicon rubber	Resistance	GF: 3.8.0.6
Strain	FDM	Polymer composite	Resistance	GF: 15
Force	FDM	Nanocomposite	Piezoresistive	—
Force	FDM	PLA	Piezoresistive	0.01–0.07 N bit^{-1}
Accelerometer	VP	Photopolymer resin	Capacitance	33.16 (fF/g^{-1})
Accelerometer	VP	Photopolymer resin	Capacitance	113 (fF/g^{-1})
Pressure	FDM	PVDF	Capacitance	16.35 Pa
Pressure	FDM	ABS	Optical absorbance	0.208 nm bar^{-1}
Pressure	Binder jetting (BJ)	Exfoliated graphite	Resistance	−0.1240.03%/°C
Temperature	Direct ink writing (DIW)	Graphene composites	Resistance	0.008%/°C
Temperature	FDM	Graphene/PLA	Electrochemical	25–100 M
Biomolecular	FDM	Biomimetic material	Electrochemical	17.7 ng cm^{-3}
Biosensor	VP	Liquid resin	Chemiluminescent	—
Biosensor	DIW	Hydrogel	Resistance	—
Cell based	FDM	PLA	Electrochemical	2 cells/ml
Cell based	VP	Photopolymer resin	Electrochemical	54 mV pH^{-1}
Microbial	MJ	Photopolymer resin	Electrochemical	—
Microbial	FDM	PLA	Electrochemical	—

type of sensor, AM process used, materials used, transduction, mechanism and sensitivity. Table 4.3 summarizes the latest information on various types of sensors, the AM process, and the materials used for sensor preparation.

4.6 Commercial availability

A market survey has been conducted to collect information on the commercial availability of sensor materials. Table 4.4, although not exhaustive, summarizes the commercial availability of materials used for printing sensors. Although new suppliers enter the market on a daily basis, the most prominent suppliers are listed in this table.

Table 4.3. 3D-printed sensors and materials.

Sensor type	AM process	Materials	Function	References
Strain sensors	FDM	PLA	Sensing components	[99]
	FDM	Graphite powder	Sensor patches	[100]
	FDM	TPU	Sensing components	[101]
	FDM	TPU	Sensing components	[102]
	FDM	Carbon black filled TPU	Flexible strain sensing	[103]
	DLP	Photo-curable resin	Porous ionogel flexible sensor	[104]
	FDM	PVDF	Piezoelectric energy harvesters	[105]
	Fused filament fabrication (FFF)	CNT and GNP filled TPU composites	Flexible strain sensors	[106]
	Aerosol-based 3D printing (Aerosol Jet® printer)	Graphene-based piezoresistive material	Strain sensors for smart tyres in autonomous vehicles	[107]
Accelerometers and displacement sensors	SLA	Resin based	Biomimetic angular acceleration sensor	[108]
	Surface-mount technology (SMD)	Nd-Fe-B magnets and Hall-effect devices	Tiny sensors to measure micro-displacements	[25]
	FDM	Polylactide material with a conductive material	Three axial piezoresistive accelerometer	[109]
Force sensors	FDM	Carbon black composite	Sensor patches	[110]
	FDM	Graphite powder	Sensor patches	[111]
Pressure sensors	FDM	CNT-PDMS	Flexible pressure sensor	[112]
	SLA	TangoPlus FLX930	Multimaterial five stretchable layers	[113]
	FDM	Carbon black (CB)/ polydimethylsiloxane (PDMS) composites	Flexible pressure sensor	[114]
Temperature sensors	FDM	PLA and graphene nano rods	Sensor capacity up to 70 °C	[115]
	IJP	Silver nanoparticle ink	Sensor capacity up to 75 °C	[116]

(Continued)

Table 4.3. (*Continued*)

Sensor type	AM process	Materials	Function	References
Particle sensors	MJ	Photopolymer	Virtual impactor for detection of particulates in the atmosphere	[117]
	VP	Resin based material	Portable particle sensor	[118]
Tactile sensors	Direct write	Multi-walled CNT-polymer composites, conductive nanocomposite + PU	Conductive tactile sensor	[119]
	VP	CNT, TangoPlus monomer	Flexible tactile sensor	[120]
	Direct-print (DP)	Rubber material	Two layers of sensing elements	[121]
	FDM	Multimaterial, and multilayer nanocomposite-based (M2A3DNC)	Biocompatible wearable tactile sensor	[122]
	Extrusion	Polyacrylamide (PAA), psyllium-N-vinylpyrrolidone (NVP), and carboxy methyl cellulose (CMC) are cross-linked using Zn^{2+} ligand	Flexible wearable sensors	[123]

Table 4.4. Materials used in the 3DP of sensors and their suppliers.

S. No.	Material and its components (generic/proprietary)	Supplier
1.	Conductive PLA: PLA + carbon black	Protopasta, Canada
2.	Clear PETG: copolyester PETG	Protopasta, Canada
4.	Metallic, high temperature/heat treating PLA (HTPLA): PLA + mica + TiO_2 + ferric oxide + colorant	Protopasta, Canada
4.	Matte fiber HPTLA: PLA + cellulose + colorant	Protopasta, Canada
5.	Copper composite fiber HTPLA: PLA + copper powder	Protopasta, Canada
6.	Composite iron fiber PLA: PLA + iron	Protopasta, Canada
7.	Composite steel fiber PLA: PLA + Fe + Ni + Mo + Si + Cr	Protopasta, Canada
8.	PLA	Sigma Aldrich, USA
9.	PI-engineering TPU (ETPU)	Rubber3dprinting.com
10.	Nickel ink, copper ink, iron ink, graphite ink, grapheme ink	Sigma Aldrich, USA

11.	Conductive graphene: PLA + graphene and graphite	BlackMagic3D, USA
12.	TPU + carbon black	Rubber3dprinting.com
13.	Electrolube silver conducting paint	Ulbrich-group, USA
14.	PE410 silver conducting ink	DuPont, USA
15.	Carbon fiber/ABS	Prusa Research, Czech Republic
16.	ABS	Sigma Aldrich, USA
17.	Commercial graphene + PLA	BlackMagic3D, USA
18.	Facilan™: PCL	3D4Makers, Netherlands
19.	Polybutylene terephthalate	DuPont, USA
20.	Stainless steel	Markforged, USA
21.	Transparent polycarbonate like material	Sigma Aldrich, USA
22.	PVDF–ceramic composites	3DXTECH, USA
23.	PET/PETG	Sigma Aldrich, USA
24.	PEEK, PEI	Nanjing Yuwei New Materials, China, and 3DXTECH, USA
25.	PVA	Sigma Aldrich, USA
26.	Tritan	Taulman, USA
27.	TPU/TPE	Sigma Aldrich, USA
28.	Thermoplastic copolyester	Sigma Aldrich, USA
29.	Nylon	Sigma Aldrich, USA
30.	Electrifi biodegradable polyester + copper	Multi3D, USA
31.	TangoPlus elastomeric polymer	Stratasys, Israel
32.	Elastomeric polyurethane	Carbon, USA
33.	Flexible 80 A elastic resin for SLA	Formlabs, USA
34.	Spot-E, a fast curing flexible resin	Spot-A Materials, US
35.	Polyimide substrate (Kapton HN) (-269 °C–400 °C)	DuPont, USA
36.	InGaN Indium gallium nitride	American Elements, USA
37.	Polymethyl methacrylate	Sigma Aldrich, USA
38.	Polyaniline	Parchem, USA
39.	Polypyrrole	Sigma Aldrich, USA
40.	PSS	Duchefa Farma, Netherlands
41.	Polyamide (PA12)	Sulptio, BASF, Germany
42.	PDMS	Sigma Aldrich, USA
43.	Objet Vero acrylic polymers	Stratasys, Israel
44.	Somos® WaterShed resins	Stratasys, Israel
45.	UV curable polymer (VisiJet M3 Crystal)	3D Systems, USA

4.7 Conclusion

It was demonstrated that 3DP is useful to print a variety of sensors, providing a simple and inexpensive way to integrate sensing and capacitive sensors, allowing for the creation of more linear sensors at the expense of increased complexity. When 3D printers are able to deposit a wider variety of materials at higher resolutions, the complexity of printable sensor systems will someday rival that of integrated circuits or perhaps even nature. Once the sensor systems of this level of complexity are joined by actuators of comparable complexity and linked with artificial intelligence, these systems may someday approach the sensing and manipulating capabilities of nature in real-world applications.

This chapter presented an extensive appraisal of some of the 3D-printed sensors for various industrial applications. Methods for creating sensors using three-dimensional printing are FDM, SLA, SLS, PJ, IJP, and DLP. This chapter mainly focused on the materials used for printing various types of sensors. Several challenges still exist in the selection of materials and their processing during the AM process. Finally, commercial materials used for printing sensors are presented.

References

[1] Dahlin A B 2012 Size matters: problems and advantages associated with highly miniaturized sensors *Sensors* **12** 3018–36

[2] Ullah F and Pun C-M 2021 The role of Internet of Things for adaptive traffic prioritization in wireless body area networks *Healthcare Paradigms in the Internet of Things Ecosystem* ed S Pal and E Balas Valentina (Amsterdam: Elsevier) pp 63–82

[3] Havskov J and Alguacil G 2004 Seismic sensors *Instrumentation in Earthquake Seismology* (Dordrecht: Springer) pp 11–76

[4] Babbar A, Sharma A, Jain V and Gupta D 2022 *Additive Manufacturing Processes in Biomedical Engineering: Advanced Fabrication Methods and Rapid Tooling Techniques* (Boca Raton, FL: CRC Press)

[5] dos Santos D M *et al* 2022 Advances in 3D printed sensors for food analysis *Trends Anal. Chem.* **154** 116672

[6] Khosravani M R and Reinicke T 2020 3D-printed sensors: current progress and future challenges *Sensors Actuators* A **305** 111916

[7] Xia F 2009 Wireless sensor technologies and applications *Sensors* **9** 8824–30

[8] Jandyal A, Chaturvedi I, Wazir I, Raina A and Ul Haq M I 2022 3D printing—a review of processes, materials and applications in industry 4.0 *Sustain. Oper. Comput.* **3** 33–42

[9] Nag A, Alahi M E E, Feng S and Mukhopadhyay S C 2019 IoT-based sensing system for phosphate detection using graphite/PDMS sensors *Sensors Actuators* A **286** 43–50

[10] Yaqoob U, Phan D-T, Uddin A I and Chung G-S 2015 Highly flexible room temperature NO_2 sensor based on MWCNTs-WO_3 nanoparticles hybrid on a PET substrate *Sensors Actuators* B **221** 760–8

[11] Alahi M E E, Pereira-Ishak N, Mukhopadhyay S C and Burkitt L 2018 An Internet-of-Things enabled smart sensing system for nitrate monitoring *IEEE Internet Things* J. **5** 4409–17

[12] Agarwal P B, Alam B, Sharma D S, Sharma S, Mandal S and Agarwal A 2018 Flexible NO_2 gas sensor based on single-walled carbon nanotubes on polytetrafluoroethylene substrates *Flex. Print. Electron.* **3** 035001

[13] Singh E, Meyyappan M and Nalwa H S 2017 Flexible graphene-based wearable gas and chemical sensors *ACS Appl. Mater. Interfaces* **9** 34544–86

[14] Tasaltin C and Basarir F 2014 Preparation of flexible VOC sensor based on carbon nanotubes and gold nanoparticles *Sensors Actuators* B **194** 173–9

[15] Da Costa T H and Choi J-W 2017 A flexible two dimensional force sensor using PDMS nanocomposite *Microelectron. Eng.* **174** 64–9

[16] Abdulla S, Mathew T L and Pullithadathil B 2015 Highly sensitive, room temperature gas sensor based on polyaniline-multiwalled carbon nanotubes (PANI/MWCNTs) nanocomposite for trace-level ammonia detection *Sensors Actuators* B **221** 1523–34

[17] Acuautla M, Bernardini S, Gallais L, Fiorido T, Patout L and Bendahan M 2014 Ozone flexible sensors fabricated by photolithography and laser ablation processes based on ZnO nanoparticles *Sensors Actuators* B **203** 602–11

[18] Zea M, Texidó R and Villa R 2021 Specially designed polyaniline/polypyrrole ink for a fully printed highly sensitive pH microsensor *ACS Appl. Mater. Interfaces* **13** 33524–35

[19] Khan S, Lorenzelli L and Dahiya R S 2014 Bendable piezoresistive sensors by screen printing MWCNT/PDMS composites on flexible substrates *Proc. of the 2014 10th Conference on Phys D. Research in Microelectronics and Electronics (Grenoble, France, 30 June)* pp 1–4

[20] Nag A, Mukhopadhyay S C and Kosel J 2017 Sensing system for salinity testing using laser-induced graphene sensors *Sensors Actuators* A **264** 107–16

[21] Woo S-J, Kong J-H, Kim D-G and Kim J-M 2014 A thin all-elastomeric capacitive pressure sensor array based on micro-contact printed elastic conductors *J. Mater. Chem.* C **2** 4415–22

[22] Muth J T *et al* 2014 Embedded 3D printing of strain sensors within highly stretchable elastomers *Adv. Mater.* **26** 6307–12

[23] Khan S, Dang W, Lorenzelli L and Dahiya R 2015 Printing of high concentration nanocomposites (MWNTs/PDMS) using 3D-printed shadow masks *Proc. of the 2015 18th AISEM Annual Conf. (Trento, Italy, 3–5 February)* pp 1–4

[24] Remaggi G, Zaccarelli A and Elviri L 2022 3D printing technologies in biosensors production: recent developments *Chemosensors* **10** 65

[25] Mattmann C, Clemens F and Tröster G 2008 Sensor for measuring strain in textile *Sensors* **8** 3719–32

[26] Yamada T *et al* 2011 A stretchable carbon nanotube strain sensor for human-motion detection *Nat. Nanotech.* **6** 296–301

[27] Zhao J, Zhang G Y and Shi D X 2013 Review of graphene-based strain sensors *Chin. Phys.* B **22** 057701

[28] Giffney T, Bejanin E, Kurian A S, Travas-Sejdic J and Aw K 2017 Highly stretchable printed strain sensors using multi-walled carbon nanotube/silicone rubber composites *Sensors Actuators* A **259** 44–9

[29] Herrmann J *et al* 2007 Nanoparticle films as sensitive strain gauges *Appl. Phys. Lett.* **91** 183105

[30] Liu H *et al* 2021 3D printed flexible strain sensors: from printing to devices and signals *Adv. Mater.* **33** e2004782

[31] Andò B, Baglio S, Lombardo C O, Marletta V and Pistorio A 2015 An inkjet printed seismic sensor *Proceedings of the IEEE Int. Instrum. Meas. Technol. Conf. (Pisa, Italy)* pp 1169–73

[32] Bodnicki M, Pakula P and Zowade M 2016 Miniature displacement sensor *Advanced Mechatronics Solutions* (Advances in Intelligent Systems and Computing, vol 393) ed R Jabłoński and T Brezina (Cham: Springer)

[33] Shi H 2015 A narrow vertical beam based structure for passive pressure measurement using two-material 3D printing *Proc. of the World Congress on Engineering and Computer Science (San Francisco, CA, USA)* pp 21–3

[34] Dankoco M D, Tesfay G Y, Benevent E and Bendahan M 2016 Temperature sensor realized by inkjet printing process on flexible substrate *Mater. Sci. Eng.* B **205** 1–5

[35] Zusman M, Schumacher C S, Gassett A J, Spalt E W, Austin E, Larson T V, Carvlin G, Seto E, Kaufman J D and Sheppard L 2020 Calibration of low-cost particulate matter sensors: model development for a multi-city epidemiological study *Environ. Int.* **134** 105329

[36] Goh G L, Yeong W Y, Altherr J, Tan J and Campolo D 2022 3D printing of soft sensors for soft gripper applications *Mater. Today* **70** 224–9

[37] Sensors *Robokits India* http://robokits.co.in/sensors (Accessed: 22 February 2023)

[38] Sensor types *AZO Sensors* https://azosensors.com/materials.aspx (Accessed: 22 February 2023)

[39] Schouten M, Wolterink G, Dijkshoorn A, Kosmas D, Stramigioli S and Krijnen G 2021 A review of extrusion-based 3D printing for the fabrication of electro- and biomechanical sensors *IEEE Sens. J.* **21** 12900–12

[40] Nassar H, Ntagios M, Navaraj W T and Dahiya R 2018 Multi-material 3D printed bendable smart sensing structures *Proc. IEEE Sens.* **2018** 3–6

[41] Guo S Z, Qiu K, Meng F, Park S H and McAlpine M C 2017 3D printed stretchable tactile sensors *Adv. Mater.* **29** 1701218

[42] Eijking B, Sanders R and Krijnen G 2017 Development of whisker inspired 3D multi-material printed flexible tactile sensors *Proc. IEEE Sens.* **2017** 1–3

[43] Christ J F, Aliheidari N, Pötschke P and Ameli A 2018 Bidirectional and stretchable piezoresistive sensors enabled by multimaterial 3D printing of carbon nanotube/thermoplastic polyurethane nanocomposites *Polymers* **11** 11

[44] Davoodi E, Fayazfar H, Liravi F, Jabari E and Toyserkani E 2020 Drop-on-demand high-speed 3D printing of flexible milled carbon fiber/silicone composite sensors for wearable biomonitoring devices *Addit. Manuf.* **32** 101016

[45] Aeby X, Dommelen R V and Briand D 2019 Fully FDM 3D printed flexible capacitive and resistive transducers *20th Int. Conf. on Solid-State Sensors, Actuators and Microsystems and Eurosensors* vol 2019 pp 2440–3

[46] Li K, Wei H, Liu W, Meng H, Zhang P and Yan C 2018 3D printed stretchable capacitive sensors for highly sensitive tactile and electrochemical sensing *Nanotechnology* **29** 185501

[47] Flowers P F, Reyes C and Ye S 2017 3D printing electronic components and circuits with conductive thermoplastic filament *Addit. Manuf.* **18** 156–63

[48] Kim M J *et al* 2019 One-step electrodeposition of copper on conductive 3D printed objects *Addit. Manuf.* **27** 318–26

[49] Saari M, Cox B, Richer E, Krueger P S and Cohen A L 2015 Fiber encapsulation additive manufacturing: an enabling technology for 3D printing of electromechanical devices and robotic components *3D Print. Addit. Manuf.* **2** 32–9

[50] Kisic M, Blaz N, Zivanov L and Damnjanovic M 2020 Elastomer based force sensor fabricated by 3D additive manufacturing *AIP Adv.* **10** 015017

[51] Wu S Y, Yang C, Hsu W and Lin L 2015 RF wireless lc tank sensors fabricated by 3D additive manufacturing *18th Int. Conf. on Solid-State Sensors, Actuators and Microsystems, Transducers 2015* vol 2015 pp 2208–11

[52] Chou B, Park J S and Kim W S 2015 3D printed inductor designs decorated with silver nano ink *2015 IEEE Nanotechnology Materials and Devices Conf.* pp 1–2

[53] Yuk H and Zhao X 2018 A new 3D printing strategy by harnessing deformation, instability, and fracture of viscoelastic inks *Adv. Mater.* **30** 1704028

[54] CCF and CBF *Anisoprint* https://anisoprint.com/product-cf (Accessed: 22 Febrauary 2023)

[55] Tully J J and Meloni G N 2020 A scientist's guide to buying a 3D printer: how to choose the right printer for your laboratory *Anal. Chem.* **92** 14853–60

[56] Ngo T D, Kashani A, Imbalzano G, Nguyen K T Q and Hui D 2018 Additive manufacturing (3D printing): a review of materials, methods, applications and challenges *Composites* B **143** 172–96

[57] Kumar R *et al* 2022 On mechanical, physical, and bioactivity characteristics of material extrusion printed polyether ether ketone *J. Mater. Eng. Perform* **32** 5885–94

[58] Silva A L, Salvador G M S, Castro S V F, Carvalho N M F and Munoz R A A A 2021 A 3D printer guide for the development and application of electrochemical cells and devices *Front. Chem.* **9** 439

[59] Bachhar N, Gudadhe A, Kumar A, Andrade P and Kumaraswamy G 2020 3D printing of semicrystalline polypropylene: towards eliminating warpage of printed objects *Bull. Mater. Sci.* **43** 171

[60] Cotabarren I and Gallo L 2020 3D printing of PVA capsular devices for modified drug delivery: design and *in vitro* dissolution studies *Drug Dev. Ind. Pharm.* **46** 1416–26

[61] El Magri A, Vanaei S and Vaudreuil S 2021 An overview on the influence of process parameters through the characteristic of 3D-printed PEEK and PEI parts *High Perform. Polym.* **33** 862–80

[62] M-base engineering + software *Material Data Center* https://materialdatacenter.com/ms/en/tradenames/Tritan/Eastman+Chemical+Co/Tritan%E2%84%A2+TX1001/72ed95a3/5701 (Accessed: 22 February 2023)

[63] Kalinke C, Neumsteir N V, Roberto de Oliveira P, Janegitz B C and Bonacin J A 2021 Sensing of L-methionine in biological samples through fully 3D-printed electrodes *Anal. Chim. Acta* **1142** 135–42

[64] Stefano J S *et al* 2022 Electrochemical (bio)sensors enabled by fused deposition modeling-based 3D printing: a guide to selecting designs, printing parameters, and post-treatment protocols *Anal. Chem.* **94** 6417–29

[65] Jain S K and Tadesse Y 2019 Fabrication of polylactide/carbon nanopowder filament using melt extrusion and filament characterization for 3D printing *Int. J. Nanosci.* **18** 1850026

[66] Podsiadły B, Skalski A, Wałpuski B and Słoma M 2019 Heterophase materials for fused filament fabrication of structural electronics *J. Mater. Sci.: Mater. Electron.* **30** 1236–45

[67] Wolterink G, Sanders R and Krijnen G 2018 Thin, flexible, capacitive force sensors based on anisotropy in 3D-printed structures *Proc. IEEE Sens.* **2018** 2–5

[68] Kwok S W *et al* 2017 Electrically conductive filament for 3D-printed circuits and sensors *Appl. Mater. Today* **9** 167–75

[69] Flandin L, Hiltner A and Baer E 2001 Interrelationship between electrical and mechanical properties of a carbon black-filled ethylene–octene elastomer *Polymer* **42** 827–38

[70] Valentine A D *et al* 2017 Hybrid 3D printing of soft electronics *Adv. Mater.* **29** 1703817

[71] Zhu Z, Wang H and Peng D 2017 Dependence of sediment suspension viscosity on solid concentration: a simple general equation *Water* **9** 474

[72] Yuk H *et al* 2020 3D printing of conducting polymers *Nat. Commun.* **11** 1604

[73] Zhuo P, Li S, Ashcroft I A and Jones A I 2021 Material extrusion additive manufacturing of continuous fibre reinforced polymer matrix composites: a review and outlook *Composites B* **224** 109143

[74] Kabir S M F, Mathur K and Seyam A-F M 2020 A critical review on 3D printed continuous fiber-reinforced composites: History, mechanism, materials and properties *Compos. Struct.* **232** 111476

[75] Mason H and Gardiner G 2020 3D printing with continuous fiber: a landscape *CompositesWorld* https://compositesworld.com/articles/3d-printing-with-continuous-fiber-a-landscape (Accessed: 22 February 2023)

[76] Espalin D, Muse D W, MacDonald E and Wicker R B 2014 3D printing multifunctionality: structures with electronics *Int. J. Adv. Manuf. Technol.* **72** 963–78

[77] Antonov F 2019 Continuous fiber 3D printing: current market review by Fedor Antonov, CEO Anisoprint *3D Printing Industry* https://3dprintingindustry.com/news/continuous-fiber-3d-printing-current-market-review-by-fedor-antonov-ceo-anisoprint-164964/

[78] Cui Y *et al* 2018 SI-traceable water content measurements in solids, bulks, and powders *Int. J. Thermophys.* **39** 1–14

[79] Kim K, Park J, Suh J, Kim M, Jeong Y and Park I 2017 3D printing of multiaxial force sensors using carbon nanotube (CNT)/thermoplastic polyurethane (TPU) filaments *Sensors Actuators A* **263** 493–500

[80] Sau K P, Chaki T K and Khastgir D 2000 The effect of compressive strain and stress on electrical conductivity of conductive rubber composites *Rubber Chem. Technol.* **73** 310–24

[81] Daalkhaijav U, Yirmibesoglu O D, Walker S and Mengüc Y 2018 Rheological modification of liquid metal for additive manufacturing of stretchable electronics *Adv. Mater. Technol.* **3** 1700351

[82] Luan C, Yao X, Liu C, Lan L and Fu J 2018 Self-monitoring continuous carbon fiber reinforced thermoplastic based on dual-material three-dimensional printing integration process *Carbon* **140** 100–11

[83] Kim H, Torres F, Wu Y, Villagran D, Lin Y and Tseng T-L 2017 Integrated 3D printing and corona poling process of PVDF piezoelectric films for pressure sensor application *Smart Mater. Struct.* **26** 085027

[84] Kim H, Torres F, Villagran D, Stewart C, Lin Y and Tseng T L B 2017 3D printing of BaTiO$_3$/PVDF composites with electric *in situ* poling for pressure sensor applications *Macromol. Mater. Eng.* **302** 1700229

[85] Bodkhe S, Rajesh P S M, Gosselin F P and Therriault D 2018 Simultaneous 3D printing and poling of PVDF and its nanocomposites *ACS Appl. Energy Mater.* **1** 2474–82

[86] Cholleti E R 2018 A review on 3D printing of piezoelectric materials *IOP Conf Ser.: Mater. Sci. Eng.* **455** 012046

[87] Kierzewski I, Bedair S S, Hanrahan B, Tsang H, Hu L and Lazarus N 2020 Adding an electroactive response to 3D printed materials: printing a piezoelectret *Addit. Manuf.* **31** 100963

[88] Iron-filled metal composite PLA *Proto-pasta* https://proto-pasta.com/products/magnetic-iron-pla (Accessed: 22 February 2023)

[89] NeoFer 25/60p *Magnetfabrik Bonn* https://magnetfabrik.de/wp-content/uploads/produkte/materiallist-according-to-din/refeb/neofer-25–60p.pdf (Accessed: 22 February 2023)

[90] Wang Z, Huber C, Hu J, He J, Suess D and Wang S X 2019 An electrodynamic energy harvester with a 3D printed magnet and optimized topology *Appl. Phys. Lett.* **114** 013902

[91] von Petersdorff-Campen K *et al* 2018 3D printing of functional assemblies with integrated polymer-bonded magnets demonstrated with a prototype of a rotary blood pump *Appl. Sci.* **8** 1275

[92] Li L *et al* 2016 Big area additive manufacturing of high performance bonded NdFeB magnets *Sci. Rep.* **6** 36212

[93] Katakam P *et al* 2015 Top-down and bottom-up approaches in 3D printing technologies for drug delivery challenges *Crit. Rev. Ther. Drug Carrier Syst.* **32** 61–87

[94] Patra S and Young V 2016 A review of 3D printing techniques and the future in biofabrication of bioprinted tissue *Cell Biochem. Biophys.* **74** 93–8

[95] Katakam P, Adiki S K and Satapathy S R 2022 Recent advancements of additive manufacturing for patient-specific drug delivery *Additive Manufacturing Processes in Biomedical Engineering* 1st edn ed A Babbar, A Sharma, V Jain and D Gupta (Boca Raton, FL: CRC Press) pp 1–26

[96] Zhang G-J and Ning Y 2012 Silicon nanowire biosensor and its applications in disease diagnostics: a review *Anal. Chim. Acta* **749** 1–15

[97] Schouten M, Wolterink G, Dijkshoorn A, Kosmas D, Stramigioli S and Krijnen G 2021 A review of extrusion-based 3D printing for the fabrication of electro- and biomechanical sensors *IEEE Sens. J.* **21** 12900–12

[98] Han T, Kundu S, Nag A and Xu Y 2019 3D printed sensors for biomedical applications: a review *Sensors* **19** 1706

[99] Yang Y, Hong C, Ahmed Abro Z A, Wang L and Yifan Z 2019 A new fiber Bragg grating sensor based circumferential strain sensor fabricated using 3D printing method *Sensors Actuators* A **295** 663–70

[100] Subramanya S B, Nagarjuna N, Prasd M G A and Nayak M M 2019 Study and tailoring of screen-printed resistive films for disposable strain gauges *Sensors Actuators* A **295** 380–95

[101] Gonzalez D, Garcia J and Newell B 2019 Electromechanical characterization of a 3D printed dielectric material for dielectric electroactive polymer actuators *Sensors Actuators* A **297** 111565

[102] Ali M M *et al* 2018 Printed strain sensor based on silver nanowire/silver flake composite on flexible and stretchable TPU substrate *Sensors Actuators* A **274** 109–15

[103] Li B, Zhang S, Zhang L, Gao Y and Xuan F 2022 Strain sensing behavior of FDM 3D printed carbon black filled TPU with periodic configurations and flexible substrates *J. Manuf. Process* **74** 283–95

[104] Peng S *et al* 2022 Tailoring of photocurable ionogel toward high resilience and low hysteresis 3D printed versatile porous flexible sensor *Chem. Eng. J.* **439** 135593

[105] Liu X, Liu J, He L, Shang Y and Zhang C 2022 3D printed piezoelectric-regulable cells with customized electromechanical response distribution for intelligent sensing *Adv. Funct. Mater.* **32** 2201274

[106] Xiang D *et al* 2020 3D printed high-performance flexible strain sensors based on carbon nanotube and graphene nanoplatelet filled polymer composites *J. Mater. Sci.* **55** 15769–86

[107] Maurya D *et al* 2020 3D printed graphene-based self-powered strain sensors for smart tires in autonomous vehicles *Nat. Commun.* **11** 5392

[108] Tiem J, Groenesteijn J, Sanders R G P and Krijnen G J M 2015 3D printed bio-inspired angular acceleration sensor *Proc. IEEE Sensors (Busan, Korea)* pp 1430–3

[109] Arh M and Slavič J 2022 Single-process 3D-printed triaxial accelerometer *Adv. Mater. Technol.* **7** 2101321

[110] Devaraj H, Yellapantula K, Stratta M, McDaid A and Aw K 2019 Embedded piezor-esistive pressure sensitive pillars from piezoresistive carbon black composites towards a soft large-strain compressive load sensor *Sensors Actuators* A **285** 645–51

[111] Nag A, Feng S, Mukhopadhyay S C, Kosel J and Inglis D 2018 3D printed mould-based graphite/PDMS sensor for low-force applications *Sensors Actuators* A **280** 525–34

[112] Gao Y, Xu M, Yu G, Tan J and Xuan F 2019 Extrusion printing of carbon nanotube-coated elastomer fiber with microstructures for flexible pressure sensors *Sensors Actuators* A **299** 111625

[113] Emon M O F, Alkadi F, Philip D G, Kim D H, Lee K C and Choi J W 2019 Multi-material 3D printing of a soft pressure sensor *Addit. Manuf.* **28** 629–38

[114] Zhu G *et al* 2022 3D printed skin-inspired flexible pressure sensor with gradient porous structure for tunable high sensitivity and wide linearity range *Adv. Mater. Technol.* **7** 2101239

[115] Sajid M, Gul J Z, Kim S W, Kim H B, Na K H and Choi K H 2018 Development of 3D-printed embedded temperature sensor for both terrestrial and aquatic environmental monitoring robots *3D Print. Addit. Manuf.* **5** 160–9

[116] Barmpakos D, Segkos A, Tsamis C and Kaltsas G 2018 A disposable inkjet printed humidity and temperature sensor fabricated on paper *Proceedings* **2** 977

[117] Zhao J, Liu M, Liang L, Wang W and Xie J 2016 Airborne particulate matter classification and concentration detection based on 3D printed virtual impactor and quartz crystal microbalance sensor *Sensors Actuators* A **238** 379–88

[118] Wang Y, Mackenzie F V, Ingenhut B and Boersma A 2018 Miniaturized 3D printed particulate matter sensor for personal monitoring *Proc. of the 17th International Meeting on Chemical Sensor (Vienna, Austria)* pp 402–3

[119] Vatani M, Engeberg E D and Choi J W 2013 Force and slip detection with direct-write compliant tactile sensors using multi-walled carbon nanotube/polymer composites *Sensors Actuators* A **195** 90–7

[120] Yun H Y, Kim H C and Lee I H 2015 Research for improved flexible tactile sensor sensitivity *J. Mech. Sci. Technol.* **29** 5133–8

[121] Vatani M, Engeberg E D and Choi J W 2015 Conformal direct-print of piezoresistive polymer/nanocomposites for compliant multi-layer tactile sensors *Addit. Manuf.* **7** 73–82

[122] Yi Q, Najafikhoshnoo S, Das P, Noh S, Hoang E, Kim T and Esfandyarpour R 2022 All-3D-printed, flexible, and hybrid wearable bioelectronic tactile sensors using biocompatible nanocomposites for health monitoring *Adv. Mater. Technol.* **7** 2101034

[123] Wu Y, Zeng Y, Chen Y, Li C, Qiu R and Liu W 2021 Photocurable 3D printing of high toughness and self-healing hydrogels for customized wearable flexible sensors *Adv. Funct. Mater.* **31** 2107202

IOP Publishing

3D Printed Smart Sensors and Energy Harvesting Devices
Concepts, fabrication and applications
Sanket Goel and Sohan Dudala

Chapter 5

CAD tools for 3D printing

Ravi Kumar Arya, Pawan Kumar, Aswin Chowdary Undavalli and Junwei Dong

The widespread adoption of 3D printing has increased the demand for reliable and efficient computer-aided design (CAD) tools that can create 3D models for 3D printing. In this chapter we present an overview of the various CAD tools used for 3D printing.

The chapter begins by introducing various computer-aided technologies, providing a foundation for understanding the various tools used in the process. The chapter then delves into the various CAD tools available, starting with the geometric modeling kernels behind them. The discussion also includes the various file types used in the different CAD tools.

The chapter concludes with a comparison of the various CAD tools, highlighting the strengths and limitations of each type of solution. The analysis provides insights into the best CAD tool options for different types of users, based on their needs, budget, and level of expertise.

5.1 Introduction

In recent years 3D printing has been in great demand for designing custom parts. 3D printing, also called additive manufacturing, can build objects with complex geometry which might not be possible with traditional manufacturing methods. It is also worth noting that 3D printing results in almost no waste material and has a comparatively short production time which results in lower costs. The automotive industry, medical/dental field, education, architecture and design, aerospace industry, jewellery, footwear, consumer home products, and art are some of the areas where 3D printing is being used and further explored. Due to its advantages, it is pertinent to explore the 3D printing techniques and the tools behind its popularity. The current trend is to use 3D printing wherever it can save time, money, or effort. The process involves creating a three-dimensional object by layering successive layers of material until the desired shape is achieved. In this chapter we will discuss computer-aided design (CAD) tools for 3D printing. CAD software has made it easier than ever to design and create intricate 3D models, which can then be printed using a 3D printer.

5.2 Computer-aided technologies

Three phases can be identified in general product development [1, 2]: the creative, conceptual, and engineering phases. Numerous tools, known as computer-aided technologies or simply 'CAx', assist in the many stages of a complete product life cycle. These tools shorten the time it takes for new items to hit the market, resulting in faster and less expensive production of products.

CAx [1–4] forms a class of different computer-based technologies to solve problems in the fields of engineering, production, medicine, architecture, etc. CAx covers a wide range of fields, including computer-aided design (CAD), computer-aided engineering (CAE), computer-aided manufacturing (CAM), computer-aided analysis (CAA), etc. Currently, various CAx fields are an integral part of product development and manufacturing. Due to their importance in this day and age, it is important to understand these technologies.

5.2.1 Computer-aided design

One of the most widely used CAx tools is computer-aided design (CAD) [5, 6] which enables designers to create and modify 2D and 3D models of products using a computer. CAD software is widely used in industries such as architecture, engineering, and manufacturing to create 2D and 3D models of products, buildings, and other structures. It allows the designers to create, modify, and analyse these models, allowing designers to make changes in real time and quickly visualize the impact of these changes. In this way, CAD has revolutionized the design process, making it faster, more efficient, and more precise than ever before. As a result, CAD has become an integral part of modern design and engineering.

Overall, CAD assists by modeling potential items virtually using computer software. Eventually, this virtual model (3D model) will be printed by the 3D printer. Many different types of CAD software are available, and each has unique capabilities that might make it ideal for a certain use.

5.2.1.1 Advantages of CAD
CAD software allows designers and engineers to create and modify designs with ease, precision, and speed. There are the following advantages to using CAD tools rather than traditional methods:

1. **Increased efficiency:** CAD software significantly reduces design time and improves design accuracy. With CAD, designers can create and modify designs quickly and easily, which reduces the time it takes to bring a product to market.
2. **Improved quality:** CAD software allows designers to create accurate and precise designs, which reduces the likelihood of errors and improves product quality. With CAD, designers can also simulate how a product will behave

under different conditions, which helps to identify potential issues before production.

3. **Time and cost savings:** CAD software can help to reduce the cost of production by minimizing the need for physical prototypes. With CAD designers can create virtual prototypes, which can be tested and refined before production. This reduces the number of physical prototypes needed, which can save time and money.

4. **Customization:** CAD software allows designers to create customized products quickly and easily. With CAD designers can modify existing designs or create new designs from scratch, which allows for greater flexibility in product design.

5. **Decreases errors:** Most CAD tools also provide tools to screen for errors in 3D models. In this way the designer is intimated about the errors beforehand which leads to fewer mistakes in the finished output.

6. **Better quality:** Not only does the design software have utility but also produces visually appealing drawings. It also offers the user a huge selection of tools to produce the drawing exactly as envisaged.

7. **CAD drawing with dimensions:** With the proper instruments and understanding of mathematical equations, even the most difficult items may be produced. This adaptability enables the designer to think creatively and develop original ideas without worrying about being unable to put them on paper. Drawings with more readability and fewer mistakes provide end products of higher caliber and accuracy.

8. **Ease of understanding:** Even the most complex designs may be easier to understand if 3D models are available to accompany with them. This cannot be done using physical sketches since it would take at least three sketches (a plan, an elevation, and a side view) to obtain a broader view. Not only is CAD fully capable of showing these perspectives of a product, it also offers the ability to interact with the object virtually to comprehend its precise parameters. Additionally, CAD models make it simpler for designers to demonstrate their work to other designers and non-engineers. In the concurrent engineering process, departments at the tail end of the entire product development process start working simultaneously while work is still being done at the first steps, and these impressive digital representations of the product can be used for marketing and sales without the need for an actual prototype.

In conclusion, CAD software has many capabilities and advantages that make it an essential tool for designers and engineers in various industries. CAD software allows designers to create accurate and precise designs, simulate real-world scenarios, collaborate with others, and create detailed documentation. With CAD, designers can significantly reduce design time, improve design accuracy, and reduce the cost of production.

5.2.2 Computer-aided engineering

Computer-aided engineering (CAE) [5] is another CAx that is used to simulate the behavior of products under different conditions. CAE software is used widely in industries such as aerospace, automotive, and manufacturing to predict how a design will perform under different conditions, such as stress, heat, and vibration. With CAE, engineers can test and optimize their designs before they are built, reducing the cost and time required for physical testing. CAE allows engineers to explore more design options, improve product performance, and reduce the risk of failure. As a result, CAE has become an essential tool for modern engineering and design. Overall, the goal of CAE tools is to offer insights into a variety of engineering analyses and aid design teams in making decisions.

5.2.3 Computer-aided manufacturing

Computer-aided manufacturing (CAM) [5, 6] uses computers to manufacture products with high precision and accuracy. CAM software is widely used in industries such as aerospace, automotive, and manufacturing to program machines such as computer numerical control (CNC) milling machines and lathes to produce parts and products with high precision and accuracy. CAM software takes the CAD models and generates the necessary tool paths and machining instructions required to manufacture the product. CAM software is widely used in the manufacturing industry to produce products with high precision and accuracy, reducing manufacturing time and costs.

5.2.4 Computer-aided analysis

Lastly, computer-aided analysis (CAA) [7] is used to analyse and optimize the performance of products. CAA software is used to simulate various operating conditions, such as heat transfer, fluid flow, and electromagnetic fields, to identify potential problems and optimize the design accordingly. CAA software is used in various industries, including aerospace, automotive, and electronics, to optimize the performance of products and reduce their environmental impact.

In general, computer-aided technologies have revolutionized the way businesses operate across various industries. These technologies have enabled faster and more efficient product development, reduced costs, and improved the quality of final products. The four key CAx tools, CAD, CAE, CAM, and CAA, have become essential tools for businesses to remain competitive in today's global economy. The continued development of these tools is expected to drive further innovation, enabling businesses to create even more sophisticated products and services.

5.3 The geometric modeling kernel

The kernel [8], also called a 'geometric modeling kernel' or 'solid modeling kernel', is the heart of a CAD system. The kernel is the code that controls how the image seen on the computer screen is defined mathematically. The kernel forms the foundation of CAD software, providing the tools necessary for creating 2D and 3D models of products, structures, and buildings. Kernels typically provide functions for creating, manipulating, and analyzing geometric objects such as points, curves, surfaces, and solids. They also provide tools for performing operations such as Boolean operations, filleting, blending, and sweeping.

Geometric modeling kernels are used in a wide range of applications, including automotive design, aerospace engineering, architecture, and product development. This is crucial because describing manufacturable shapes in a mathematical representation involves making choices about how each shape is calculated and stored. Thus kernels help choose how to calculate and store each shape when defining manufacturable shapes in a mathematical representation. In this way, a kernel facilitates the usage of a CAD system and even makes it possible.

However, there are different kernels on the market from various vendors and these kernels can have different representations for the same shape. The accuracy, efficiency, and robustness of a kernel can greatly affect the performance and quality of CAD software. These kernels provide a range of features and capabilities, including support for different mathematical representations of geometry, different data structures, and different file formats. Each kernel has its strengths, weaknesses, and peculiarities and it is important to understand the differences among different kernels. In the following section, we discuss the major kernels that dominate the CAD market.

5.3.1 ACIS

ACIS (Advanced Computerized Implementation of Standards) [9–13] is a popular commercial geometric modeling kernel which was developed by Spatial Corporation, now part of Dassault Systèmes. It is known for its robustness and ability to handle complex geometry. Its popularity can be seen in the fact that a lot of CAD modeling packages rely on it completely.

ACIS is based on the boundary representation (often abbreviated to BREP or B-rep) data structure, which is a popular way to represent solid geometry in CAD systems. The BREP data structure provides a way to describe the geometry of a solid object by breaking it down into simpler parts, such as faces, edges, and vertices. This allows designers to create and manipulate complex solid models with ease.

ACIS provides a wide range of modeling capabilities, including the ability to create and manipulate both 2D and 3D models. It supports a range of geometric primitives, such as points, lines, curves, and surfaces, which can be combined to create more complex shapes. ACIS also provides advanced features such as filleting, chamfering, and blending, which allow designers to create smooth transitions between different shapes and surfaces.

ACIS supports two kinds of data files, Standard ACIS Text (SAT) with a .sat file extension, and Standard ACIS Binary (SAB) with a .sab file extension. The SAT file is an ASCII text file so can be viewed with a simple text editor while the SAB file is meant for compactness and not for human readability, making it difficult to view with a simple text editor.

ACIS is used by several famous software products, including Ansys, AutoCAD, Inventor, SolidWorks, BricsCAD, SpaceClaim, TurboCAD, and Cimatron.

5.3.2 Parasolid

Parasolid [14] is a commercial 3D geometric modeling kernel from Siemens Digital Industries Software. Parasolid is also based on BREP and supports a range of modeling techniques, including solid modeling, direct editing, and free-form surface/sheet modeling. Parasolid provides wide-ranging graphical and rendering support, including precise hidden line and wireframe, as well as versatile tessellation functionality and a full suite of model data inquiries. Overall, Parasolid is a reliable and powerful kernel that can handle complex geometries and provide high-quality CAD models.

Parasolid has two primary file extensions, ASCII with the .x_t file extension, and binary with the .x_b file extension.

Parasolid has been integrated into several popular CAD software products, including SolidWorks, COMSOL Multiphysics, Ansys, Altair HyperWorks, and Siemens NX.

5.3.3 Open CASCADE technology

Open CASCADE Technology (OCCT) [15] is an open-source software development platform for creating 3D modeling and simulation applications. It was developed by Open Cascade, a subsidiary of Capgemini. OCCT is widely used in industries such as automotive, aerospace, and mechanical engineering.

OCCT is designed as a comprehensive suite of tools for 3D modeling, simulation, and visualization. It includes a range of advanced features for creating and manipulating 3D models, such as support for parametric modeling, meshing, and visualization. The platform also includes tools for creating and managing geometric data, such as curves and surfaces, as well as algorithms for detecting collisions and computing intersections.

The major key advantage of OCCT is its open-source nature. As an open-source platform, OCCT is freely available for anyone to use and modify, making it a popular choice for developers who want to create custom 3D modeling and simulation applications. The platform is also actively developed and maintained by a large community of developers, ensuring that it remains up-to-date with the latest advances in 3D modeling and simulation technology.

Overall, OCCT is a powerful and flexible platform for creating 3D modeling and simulation applications. Its modular design, support for a wide range of 3D file formats, and advanced visualization tools make it a popular choice for developers working in industries such as automotive, aerospace, and mechanical engineering.

OCCT is used in several popular CAD software products such as FreeCAD, KiCad, SALOME, Gmsh, etc.

Indeed, geometric modeling kernels are essential components of CAD software, providing the underlying algorithms and data structures necessary to create, manipulate, and analyse geometric models. These kernels enable designers to create complex and accurate models of products, structures, and buildings efficiently. With the continued development of these kernels, we can expect to see even more sophisticated and complex models in the future.

5.4 3D printing process

In a typical 3D printing process, first the 3D model at hand is sliced into multiple layers. Then the structure is manufactured layer by layer (bottom-up) with the help of a 3D printer to create the whole desired structure. While the whole 3D printing process might appear easy, it involves several steps to manufacture the desired shape.

For simplicity, the 3D printing software pipeline can be explained with the help of figure 5.1 [16].

The whole 3D printing process can be divided into the following steps:

1. **Input model:** The first step is to create a printable digital model for the 3D printer. Such a 3D printable model is normally created using CAD software, a 3D scanner, or digital cameras, or can be downloaded directly from several websites [17, 18]. Normally, the stereolithography file format (STL) is used as the file format to store the input model files. For other popular file formats, check the next section.

2. **Orientation and positioning:** The second step is to orient and position the digital model in the printing software. This involves selecting the orientation of the model and its position on the printing bed. The orientation and positioning of the model can have a significant impact on the quality of the final product and the time it takes to print.

3. **Support structure determination:** The third step is to determine the need for support structures. Support structures are structures that are printed along with the model to provide support and stability during the printing process. They are usually necessary for complex designs or designs with overhangs. The support structures are generated automatically by the printing software. It helps to support overhangs, maintain stability, and prevent curling as materials harden.

4. **Slicing:** In simple language, slicing is a process that converts a 3D model into a language that a 3D printer understands. The slicer or slicing software cuts a model into flat layers which a 3D printer can print layer by layer. The slicer instructs the printer in a language called G-code with instructions such as resolution, printing speed, or layer height. For FDM 3D printers, there are two categories of slicers: universal open-source software such as Cura (developed by Ultimaker), Repetier, and Slic3r and paid software such as Simplify3D but also proprietary software such as ReplicatorG at MakerBot, ZSuite at Zortrax, and Voxelizer at ZMorph.

5. **Path planning:** The fifth step is to plan the path that the printer head will follow to create the object. This involves determining the order in which the layers will be printed and the path that the printer head will follow for each layer. The printing software calculates the path for the printer head based on

Input Model

⬇

Orientation and Positioning

⬇

Support Structure Determination

⬇

Slicing

⬇

Path Planning

⬇

Machine Instructions

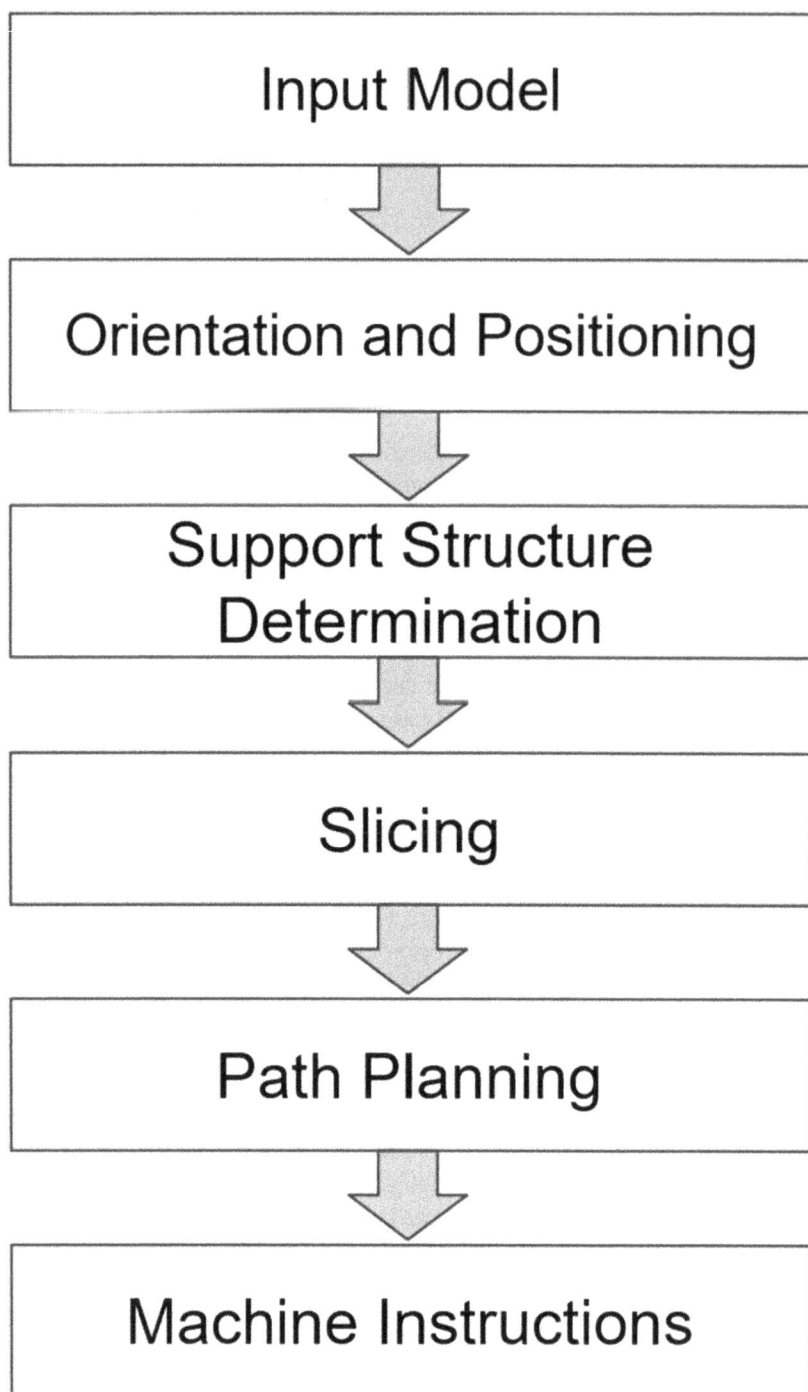

Figure 5.1. 3D printing software pipeline.

the design and the selected printing parameters. Paths affect the build time, surface accuracy, stiffness, strength, and post-manufacture distortion. Path planning helps to optimize the build time, increase surface accuracy, and regulate distortion, stiffness, and strength.

6. **Machine instructions:** The final step is to generate the machine instructions that will control the printer. This involves converting the path plan into machine code that the printer can understand. The machine code is then sent to the printer, which uses it to create the object layer by layer.

In short, the 3D printing process involves inputting a digital model, orienting and positioning it, determining the need for support structures, slicing it into thin layers, planning the path, and generating the machine instructions that will control the printer. Each step is critical to the success of the printing process, and careful consideration must be given to each one to ensure that the final product meets the desired specifications.

5.5 File formats

There are several file formats used in 3D printing, and each has its advantages and disadvantages. Here is a brief overview of the most common file formats used in 3D printing:

1. **STL (stereolithography)**
   STL is one of the most important neutral 3D file formats in the domain of 3D printing. STL encodes an approximate surface geometry of a 3D model using a triangular mesh. STL only stores the geometry of the model and, therefore, is one of the simplest and leanest 3D file formats available. The STL format is specified in ASCII and binary representations, out of which the binary representation creates compact STL files. The STL file format has the extension .stl.

2. **OBJ (object)**
   The OBJ file format is another neutral heavyweight in the field of 3D printing. It is also widely used in 3D graphics. It is a simple text-based format that defines the geometry of a 3D object, including its vertices, faces, and texture coordinates. OBJ files can also store information about the material and color of the model, making it a useful format for creating realistic and detailed 3D prints. The 3D file format has the extension .obj.

3. **AMF (additive manufacturing file format)**
   AMF is a newer file format designed specifically for 3D printing, and it aims to improve on the limitations of STL files. AMF files can support higher resolution and more complex geometry than STL files, and they can also include information about materials, color, and texture. Therefore, the AMF file format is used in 3D printers where printing of multiple colors is needed. However, the format is not yet widely supported by 3D printing software and hardware.

4. **3MF (3D manufacturing format)**
   3MF is another newer file format designed for 3D printing, and it aims to be a more comprehensive and streamlined alternative to STL. 3MF files can include information about geometry, color, texture, materials, and print settings, and they are designed to be easy to work with and share across different software applications and devices.

In summary, STL and OBJ are older file formats that have been widely used in 3D printing, while AMF and 3MF are newer formats that aim to improve on the limitations of older formats. The choice of file format will depend on the specific requirements of the 3D printing project, including the complexity of the model, the desired surface quality, and the available software and hardware.

5.6 CAD tools

There are numerous CAD tools available in the market, and each one has its strengths and weaknesses. Here we compare the most well-known CAD tools (in alphabetical order) prevalent in the 3D printing industry in terms of their interface, license, pricing, major supported file types, and their advantages and disadvantages.

1. **Tool name:** 3ds Max
   Developer(s): Autodesk
   Interface: Software
   License: Proprietary
   Pricing: Starting at $1,875/year
   Major supported file types: 3DS, DXF, FBX, IGES, SAT, STEP, STL, VRML
   Advantages: A powerful and versatile 3D modeling tool with advanced features for animation, rendering, and simulation; with a large community of users; widely used in the gaming and entertainment industries.
   Disadvantages: Can be expensive for small businesses and freelancers; an overwhelming interface for beginners; requires a high-performance computer to run smoothly.
   Website: https://www.autodesk.com/products/3ds-max/overview

2. **Tool name:** AC3D
   Developer(s): Inivis Limited
   Interface: Software
   License: Proprietary
   Pricing: Starting at $179 per unit
   Major supported file types: 3DS, DXF, OBJ, STL
   Advantages: A lightweight and affordable 3D modeling tool; easy to use; offers features for UV mapping, texture painting, and animation; its files can be exported to a wide range of formats.
   Disadvantages: It may not be suitable for complex designs or large-scale projects; may also lack some advanced features compared to other CAD tools.

Website: http://www.inivis.com

3. **Tool name:** AutoCAD 3D

    Developer(s): Autodesk

    Interface: Software

    License: Proprietary

    Pricing: Starts at $1,955/year

    Major supported file types: 3DS, DWG, DXB, DGN, PDF, SAT

    Advantages: A widely used CAD tool for 3D modeling, drafting, and documentation; offers advanced features for parametric modeling, assembly design, and simulation; a wide range of file format support.

    Disadvantages: Can be expensive for small businesses and freelancers; an overwhelming interface for beginners.

    Website: https://www.autodesk.com/products/autocad/overview

4. **Tool name:** Blender

    Developer(s): Blender Foundation, community

    Interface: Software

    License: Open source

    Pricing: Free

    Major supported file types: FBX, OBJ, PLY, STL

    Advantages: A free and open-source 3D modeling tool with advanced features for animation, rendering, and simulation; has a large community of users and is widely used in the gaming and entertainment industries.

    Disadvantages: The interface can be complex for beginners; it may also have limitations on complex designs or large assemblies due to performance issues.

    Website: https://www.blender.org

5. **Tool name:** BricsCAD

    Developer(s): Bricsys

    Interface: Software

    License: Proprietary

    Pricing: $1,265 one-time price for BricsCAD Pro

    Major supported file types: IGES, PDF, SAT, STEP, STL, X_T

    Advantages: More affordable than many other options; high compatibility with other file formats.

    Disadvantages: Limited community support; no macOS support; limited industry-specific tools.

    Website: https://www.bricsys.com

6. **Tool name:** BRL-CAD

    Developer(s): Army Research Laboratory

    Interface: Software

    License: Open source

    Pricing: Free

    Major supported file types: DXF, IGES, OBJ, STL, VRML

Advantages: Large community; supports multiple file formats; advanced capabilities such as ray tracing, global illumination, and physics-based simulations.

Disadvantages: Steep learning curve for beginners; limited graphical user interface; has performance issues for large models.

Website: https://brlcad.org

7. **Tool name:** Creo

Developer(s): PTC

Interface: Software

License: Proprietary

Pricing: Starts at $2,780/year

Major supported file types: IGES, STEP, and STL

Advantages: A powerful CAD tool that offers advanced features for parametric modeling, assembly design, and simulation; has a large community of users and is widely used in the manufacturing industry.

Disadvantages: It can be expensive, especially for small businesses and freelancers; can be an overwhelming interface for beginners.

Website: https://www.ptc.com/en/products/creo/parametric

8. **Tool name:** DesignSpark Mechanical

Developer(s): Ansys, Inc, RS Components

Interface: Software

License: Proprietary

Pricing: Free

Major supported file types: IGES, OBJ, STEP, STL

Advantages: Easy to use; direct modeling; large component library; multiple export options.

Disadvantages: Limited advanced features; limited import options; limited collaboration options; only available for Windows OS.

Website: https://www.rs-online.com/designspark/mechanical-software

9. **Tool name:** FreeCAD

Developer(s): Open-source community

Interface: Software

License: Open source

Pricing: Free

Major supported file types: DAE, DXF, IFC, IGES, IV, OBJ, SCAD, STEP, STL

Advantages: Open source; free; supports multiple file types; powerful parametric modeling capabilities; extensible through plugins; active community support.

Disadvantages: Steep learning curve; limited documentation; the interface can be overwhelming for new users.

Website: https://www.freecad.org

10. **Tool name:** Fusion 360

Vendor: Autodesk

Interface: Software

License: Proprietary

Pricing: Starts at $545/year

Major supported file types: DWG, DXF, IGES, OBJ, STEP, STL

Advantages: Comprehensive design and engineering toolset; cloud-based collaboration and data management; supports CNC and 3D printing workflows; integrated simulation and CAM capabilities.

Disadvantages: Subscription-based pricing model; requires an Internet connection; limited file format support; learning curve for advanced features.

Website: https://www.autodesk.com/products/fusion-360/overview

11. **Tool name:** IronCAD

    Developer(s): IronCAD LLC

    Interface: Software

    License: Proprietary

    Pricing: Starts at $137 for the trial version

    Major supported file types: 3DS, DWG, DXF, OBJ, IGES, STEP, STL, X_T

    Advantages: Intuitive interface; the hybrid modeling approach allows for both direct and parametric modeling; flexible licensing options; integrated visualization; rendering capabilities.

    Disadvantages: Limited simulation and analysis capabilities; limited community support; higher cost compared to some other options.

    Website: https://www.ironcad.com

12. **Tool name:** Inventor

    Developer(s): Autodesk

    Interface: Software

    License: Proprietary

    Pricing: Starts at $2,415/year

    Major supported file types: DWG, DXF, IGES, OBJ, STEP, STL

    Advantages: Comprehensive design and engineering toolset; supports advanced simulations and animations; integrated CAM capabilities; cloud-based collaboration and data management.

    Disadvantages: Expensive subscription-based pricing model; requires an Internet connection; steep learning curve for advanced features.

    Website: https://www.autodesk.com/products/inventor/overview

13. **Tool name:** LeoCAD

    Developer(s): LeoCAD Development Team

    Interface: Software

    License: Open source

    Pricing: Free

    Major supported file types: LDR, MPD

    Advantages: Open source and free; designed specifically for building with LEGO bricks; extensive library of LEGO parts; supports building instructions and animations.

Disadvantages: Limited to building with LEGO bricks; limited community support; may not be suitable for non-LEGO projects; limited file types supported.

Website: https://www.leocad.org

14. **Tool name:** Maya

    Developer(s): Autodesk, Inc.

    Interface: Software

    License: Subscription-based

    Pricing: Starts at $1,875/year

    Major supported file types: DXF, DWG, EPS, IGES, OBJ, STEP, STL, TIF

    Advantages: A comprehensive toolset for 3D animation, modeling, and rendering; integrated simulation and effects capabilities; supports complex character rigging and animation; widely used in film and game industries.

    Disadvantages: Expensive subscription-based pricing model; a steep learning curve for advanced features; requires powerful hardware to run smoothly.

    Website: https://www.autodesk.com/products/maya/overview

15. **Tool name:** Onshape Free

    Developer(s): PTC

    Interface: Web-based

    License: Proprietary

    Pricing: Free, paid license starts at $1,500/year

    Major supported file types: 3MF, IGES, OBJ, STEP, STL, X_T

    Advantages: Good collaboration capability; a wide range of learning resources.

    Disadvantages: Limited storage/features for the free plan; Internet connection required for the free plan.

    Website: http://onshape.com

16. **Tool name:** OpenSCAD

    Developer(s): Marius Kintel, Claire Wolf

    Interface: Software

    License: Open source

    Pricing: Free

    Major supported file types: AMF, OFF, STL

    Advantages: Good for parametric modeling and creating precise designs using code.

    Disadvantages: Requires knowledge of programming and mathematical concepts.

    Website: http://openscad.org

17. **Tool name:** Rhinoceros 3D

    Developer(s): Robert McNeel and Associates

    Interface: Software

    License: Proprietary

    Pricing: Starts at $995 (commercial)

Major supported file types: 3DS, DWG, DXF, EPS, IGES, OBJ, SAT, STEP, STL, VRML

Advantages: Suitable for complex designs and precise modeling.

Disadvantages: Costly.

Website: https://www.rhino3d.com

18. **Tool name:** SketchUp

   Developer(s): Trimble Inc

   Interface: Software and web-based

   License: Proprietary

   Pricing: Free (basic version), $119/year (Go version)

   Major supported file types: 3DS, DXF, DWG, OBJ, PDF, SKP, STL, VRML

   Advantages: User-friendly interface; suitable for quick designs and basic modeling.

   Disadvantages: Limited in terms of precise modeling and complex designs.

   Website: https://www.sketchup.com

19. **Tool name:** SolidWorks

   Developer(s): Dassault Systèmes SE

   Interface: Software

   License: Proprietary

   Pricing: Starts at $99/year

   Major supported file types: 3DM, 3MF, DXF, DWG, IGES, OBJ, PDF, SAT, STEP, STL

   Advantages: Suitable for complex designs and precise modeling.

   Disadvantages: High cost.

   Website: https://www.solidworks.com

20. **Tool name:** TinkerCAD

   Developer(s): Autodesk

   Interface: Web-based

   License: Proprietary

   Pricing: Free

   Major supported file types: OBJ, STL, SVG

   Advantages: User-friendly interface; suitable for basic modeling and quick designs.

   Disadvantages: Limited in terms of precise modeling and complex designs.

   Website: https://www.tinkercad.com

21. **Tool name:** TurboCAD

   Developer(s): IMSI/Design, LLC

   Interface: Software

   License: Proprietary

   Pricing: Starts at $99.99/year

   Major supported file types: 3DM, 3DS, DWG, DXF, IGS, OBJ, SAT, SKP, STEP, STL, TIF

       Advantages: Offers a range of 2D and 3D design tools.
       Disadvantages: Limited parametric modeling capabilities.
       Website: https://www.turbocad.com
22. **Tool name:** Wings3D
       Developer(s): Björn Gustavsson, Dan Gudmundsson, and others
       Interface: Software
       License: Open source
       Pricing: Free
       Major supported file types: 3DS, OBJ, STL, SVG, X3D
       Advantages: User-friendly interface.
       Disadvantages: Limited in terms of advanced features compared to other CAD tools.
       Website: http://www.wings3d.com

It is pertinent to mention that the rudimentary use of these typical tools might be taught in under an hour [19] but complex functions have a steeper learning curve. It helps to start the user quickly to experience the charm of 3D printing and get them hooked on it. Most of these software products allow parametric designs, which help modify the models to suit user needs.

If a user wants to go from 'zero to plastic' as soon as possible, it will be worth trying Tinkercad and OpenSCAD, both of which are free.

5.7 Conclusion

In conclusion, CAD tools for 3D printing have revolutionized the way designers create 3D models. The software makes it easy for designers to create accurate and complex models quickly, which can then be printed using a 3D printer. There are several CAD tools available for 3D printing, ranging from beginner-friendly software to more advanced tools used by professionals. CAD tools have several benefits, including ease of use, speed, accuracy, and collaboration. As 3D printing continues to evolve, we can expect to see more advanced CAD tools that will make it easier than ever to create intricate 3D models.

References

[1] Dankwort C W, Weidlich R, Guenther B and Blaurock J E 2004 Engineers' CAx education —it's not only CAD *Comput. Aided Des.* **36** 1439–50
[2] Łukaszewicz A, Skorulski G and Szczebiot R 2018 The main aspects of training in the field of computer-aided techniques (CAx) in mechanical engineering *Proc. of 17th Int. Scientific Conf. on Engineering for Rural Development* pp 865–70
[3] 2016 *Computer-aided Technologies—Applications in Engineering and Medicine* ed R Udroiu (London: IntechOpen)
[4] Li Y, Hedlind M and Kjellberg T 2015 Usability evaluation of CADCAM: state of the art *Procedia CIRP* **36** 205–10
[5] Lee K 1999 *Principles of CAD/CAM/CAE Systems* (Boston, MA: Addison-Wesley Longman)

[6] Narayan K L 2008 *Computer Aided Design and Manufacturing* (New Delhi: Prentice Hall of India)

[7] Nunamaker Jr J F, Konsynski Jr B R, Ho T and Singer C A 1976 Computer-aided analysis and design of information systems *Commun. ACM* **19** 674–87

[8] Stevenson J 2013 GrabCAD tips: the kernel, why CAD systems don't play well with others *GrabCAD Blog* https://web.archive.org/web/20220925014340/https://blog.grabcad.com/blog/2013/05/14/kernels-why-cad-systems-dont-play-well-with-others/

[9] CAD data interoperability around ACIS neutral format *CAD Interop* https://cadinterop.com/en/formats/neutral-format/acis.html

[10] 3D ACIS Modeler *Spatial Corp* https://spatial.com/products/3d-acis-modeling

[11] Corney J R and Lim T 2001 *3D Modeling with the ACIS Toolkit* (Kippen: Saxe-Coburg)

[12] 3D ACIS Modeler (Archived) *Spatial Corp* https://web.archive.org/web/20070309194852/http://spatial.com/products/acis.html

[13] Yang X, Yu W and Mittra R 2005 Conformal FDTD package development based on ACIS and HOOPS *2005 Asia-Pacific Microwave Conf. Proc.* vol 5 (Piscataway, NJ: IEEE) p 4

[14] Parasolid *Siemens* https://plm.automation.siemens.com/global/en/products/plm-components/parasolid.html

[15] The OpenCASCADE Technology http://opencascade.org/

[16] Shamir A, Bickel B and Matusik W 2016 Computational tools for 3D printing *ACM SIGGRAPH* Course Program https://web.archive.org/web/20220601191753/http://computational-fabrication.com/

[17] UltiMaker Thingiverse www.thingiverse.com

[18] You Magine www.youmagine.com

[19] Schelly C, Anzalone G, Wijnen B and Pearce J M 2015 Open-source 3-D printing technologies for education: bringing additive manufacturing to the classroom *J. Visual Lang. Comput.* **28** 226–37

IOP Publishing

3D Printed Smart Sensors and Energy Harvesting Devices
Concepts, fabrication and applications
Sanket Goel and Sohan Dudala

Chapter 6

3D-printed composite sensors: advancements, opportunities, and prospects

Arnab Mukherjee, Adil Wazeer, Apurba Das, Arijit Sinha and Shrikant Vidya

In recent years 3D printing technology has been used extensively in the production of numerous sophisticated components and parts in different areas of research and development. Over the last few decades, electronic gadgets have become commonplace, and with them, the necessity for sensors. Although there are several ways to make sensors, 3D printing has become increasingly popular recently because of the sensors' need for absolute precision and dimensional accuracy. 3D printing, or additive manufacturing (AM), utilizes the melting and solidification of filler materials in the fabrication of components. This method is being improved to create high-performing composites by combining various components in the filler material to display various desired features. Biomedical, mechanical, and electrically optimised products currently employ 3D-printed composites. In this chapter we examine the modification of electrical components—primarily sensors—using 3D printing composites. The creation of 3D-printed composites, their advantages over traditional manufacturing techniques, the challenges encountered, and their prospects for the future are all discussed thoroughly.

6.1 Introduction

3D printing technology plays a major role in production technologies for which various new advanced and complex manufacturing processes have been developed for over a decade. Additive manufacturing or 3D printing has become an important part of the manufacturing industry due to its wide range of applications and ease-of-use [1–5]. The 3D printing method allows the manufacturing of components layer by layer, unlike traditional hand assembly of each of the parts, resulting in the manufacturing of complex shapes [6]. The layers designed using computer assisted design (CAD) software are sent to the 3D printer which can construct the layers in various ways. The powder method, laser sintering, electron beam, and lithographic process are various methods of constructing the 3D product [7–11].

Recently this 3D printing technology has also been used in the manufacture of various electronic components and sensors [12, 13]. Sensors are used in the measurement of important properties in engineering, such as force, displacement, pressure, and strain. Sensors need to provide accurate and precise values and so they are manufactured with great care and new manufacturing methods have been tried and tested over the years, such as lamination, lithography, coating, and paper-based sensors [14, 15]. Using 3D printing technology for electronic sensors has gained interest due to its low cost, accuracy, and fast fabrication [16, 17], unlike the high cost and time-consuming nature of other manufacturing processes [18]. Studies on the electrical, mechanical, and piezoresistive behaviour of 3D fabricated sensors have shown that it is a favourable manufacturing process for sensors which can be customized according to different specifications and needs. Low human intervention and high flexibility also makes it a more favourable manufacturing process.

Due to the good adhesion between the thermoplastic polyurethane layers in the fabrication of highly elastic strain sensors, little to no degradation was observed [19]. 3D-printed carbo-morph material has been used in determining the changing resistivity of mechanical stress due to its piezoresistive behaviour [20]. Using carbon nanotubes (CNTs) as electrically conducting composite elastomers, various piezoresistive sensors and capacitive sensors have been fabricated using carbon black incorporating it into 3D fused deposition modelling (FDM) printable resins [21].

6.2 3D printing technologies

3D printing (or additive manufacturing) is being used widely because of its decreased energy consumption and ability to produce complex components in a short amount of time [22]. 3D printing technology can be divided into several categories: binder jetting (BJ), material extrusion (ME), directed energy deposition (DED), material jetting (MJ), sheet lamination (SL), powder bed fusion (PBF), and vat photopolymerization (VP) [23]. Each of the processes has its own set of advantages and disadvantages and is selected and used accordingly.

Additive manufacturing is the preferred choice in manufacturing electrical components compared to traditional casting processes for several important reasons, starting with its simplicity. To print 3D-printed parts, we just need to design a 3D model, unlike the need to create a mould in the metal casting process, which saves a lot of time and effort. In the post-production of cast metal parts, the metal removal process involves machining off the unwanted parts, such as the filling system, and also the manual removal of metal that was spilled between the moulds, followed by a surface finishing process. In the 3D printing process, only the support needs to be removed after the production process, making it a feasible choice for manufacturing.

6.3 Sensor fabrication

One or many 3D printing technologies can be used in the fabrication of a sensor. Sensors are mainly manufactured for two fields, engineering and medicine, and can be further classified into many other types of sensors with different applications. Improving the manufacturing techniques of electronic device has led to the additive manufacturing of sensors. Inkjet-printed circuits were one of the first 3D-printed components used in an electronic device [24].

The fabrication of sensors using additive manufacturing typically involves the following procedure:

1. **Design:** The first step is to design the sensor using CAD software. The design should be optimized for additive manufacturing and should include any necessary support structures.
2. **Material selection:** Select the appropriate materials for your sensor application. This may include selecting conductive and non-conductive materials.
3. **Printing:** Once the design is finalized, it is time to print the sensor using additive manufacturing techniques such as 3D printing. The printing process should be controlled carefully to ensure the accuracy and precision of the sensor.
4. **Post-processing:** After printing, the sensor may require post-processing such as annealing or sintering to improve its properties.
6. **Integration:** The printed sensor can now be integrated with any necessary electronics and other components to create a functional sensor system.
6. **Testing:** The final step is to test the sensor to ensure it meets the necessary specifications and performance requirements.

It is important to note that the specific details of the fabrication procedure may vary depending on the type of sensor being produced, the materials used, and the additive manufacturing technique employed.

Additive manufacturing has several advantages over traditional manufacturing processes when it comes to sensor fabrication. Here are some key differences:

1. **Design flexibility:** With additive manufacturing, it is possible to create complex geometries and structures that are difficult or impossible to produce using traditional manufacturing techniques. This allows for more customized and optimized designs for specific sensor applications.
2. **Reduced material waste:** Additive manufacturing only uses the materials needed to create the final product, reducing material waste and cost.
3. **Improved speed and efficiency:** Additive manufacturing can produce parts more quickly and efficiently than traditional manufacturing methods, allowing for faster turnaround times and lower costs.
4. **Reduced tooling costs:** Traditional manufacturing often requires expensive tooling, moulds, and fixtures, which can be a significant cost for small batch or customized production runs. Additive manufacturing eliminates the need for these tools, reducing costs and lead times.
6. **Material selection:** Additive manufacturing allows for the use of a wider range of materials, including metals, polymers, and composites, which can improve the performance and functionality of sensors.

However, there are also some limitations to additive manufacturing for sensor fabrication. For example, additive manufacturing may not be suitable for high-volume production runs, and the quality of the finished product may be affected by factors such as layer thickness, printing orientation, and post-processing steps. Additionally, the cost of the equipment and materials required for additive manufacturing can be high compared to traditional manufacturing techniques.

Other manufacturing processes can be used for sensor fabrication as well as additive manufacturing. There are several traditional manufacturing processes that can be used for sensor fabrication. Here are some examples:

1. **Photolithography:** This process involves using a photoresist material and a mask to transfer a pattern onto a substrate. The patterned substrate is then etched to create the sensor structure.
2. **Thin film deposition:** This process involves depositing a thin film of a material onto a substrate using techniques such as sputtering, evaporation, or chemical vapour deposition. The deposited material can be patterned using a mask or other techniques.
3. **Electroforming:** This process involves using an electrodeposition process to create a metal structure on a mould or substrate. The mould or substrate is coated with a conductive material and then immersed in an electrolytic bath to deposit the metal.
4. **CNC machining:** This process involves using computer-controlled machines to cut or shape materials to create the sensor structure. CNC machining can be used to manufacture the structure of the sensor. Injection moulding is often used for high-volume production runs. The range of materials includes metals, plastics, and composites.
6. **Injection molding:** This process involves injecting a molten material into a mould to create the sensor structure. Injection molding is often used for high-volume production runs.

These traditional manufacturing processes have their own advantages and disadvantages compared to additive manufacturing. For example, photolithography and thin film deposition can create high precision structures, but may not be suitable for complex geometries or large-scale production runs. CNC machining and injection molding can create complex geometries and are suitable for high-volume production runs, but may be more expensive for small batch runs or customized designs.

6.3.1 Engineering sensors

Engineering sensors are subdivided into mechanical, tactile, temperature, and particle sensors. Parameters such as force, pressure, acceleration, displacement, and strain can be measured through the application of mechanical 3D-printed sensors [25–27]. Newly emerging 3D printing technology has even enabled sensors to be fitted into textiles [28]. Strain sensors or strain gauges detect tensile and compressive loads and convert them into electrical signals [29–31]. Strain sensors can be customised into required shapes using soft matrices [32]. Graphene, nanotubes, or nanoparticles have been utilized for the fabrication of strain sensors [33–36]. Carbon nanotubes alone are not a good choice of material for highly sensitive strain sensors [37]. Miniature sensors were fabricated using 3D printing technology to measure micro displacements [38]. A six-sided gaming die with an accelerometer and displacement sensors was made using a 3D printing process capable of detecting motion and top surface recognition via gravity [39]. In the case of force sensors, the applied forces are converted into electronic signals. Force sensors are widely used in almost all applications ranging from medical instruments to automotive devices and robotics [40–43]. Force sensors mainly consist of a transducer, a flexure, and their housing. Commercial force sensors have traditionally been costly and fragile, and thus AM has been used to create affordable, easily customizable force sensors [44]. Piezoelectric transducers or strain gauges are utilised to give an output signal in the most popular force sensors. Consequently, complicated signal conditioning is required. The installation of 3D-printed transducers that are suitable for a variety of flexure dimensions and displacements is simple [45]. Different pressure sensors can identify and monitor pressure variations, structural stresses, and a goal pressure. 3D printing can be used to create a pressure sensor, as discussed in [46]. In order to monitor a target pressure, a beam-based structure was manufactured using the FDM technique. By altering the printed part's geometric specifications, various pressures could be measured. A printed pressure sensor was created concurrently [47] for use in soft robotics and wearable electronics. Polydimethylsiloxane (PDM) material was utilised in this situation to provide the device flexibility. A combination of an uncured elastomer and PDM was used in the ME printing process to build the sensor. The electrodes of the sensors were been made of printed micro-structures. The manufactured sensors were employed for the purpose of detecting pressing, twisting, and bending forces. Based on the findings, it has been reported that the printed sensors had great sensitivity and long-term dynamic stability. More recently, a multi-material 3D print made up of five different stretchable layers has been used [48].

The creation of temperature sensors may be done via 3D printing, much as for other forms of sensors. The authors of [49] report on the creation of an inkjet-printed temperature sensor. A micrometre-sized temperature sensor was later created using 3D printing technology in [50]. In this case, a photoresist based on a photolithographic technique was combined with nanoparticles. In brief, the researchers employed a two-photon polymerization-based 3D direct laser writing approach. This 3D printing method allows for the production of micrometre-scale structures. The constructed sensor was utilised on an electrical chip to measure the temperature. The authors of [51] describe the use of inkjet-based 3D printing to create a flexible temperature sensor. Composed silver was placed on a polymeric substrate to create the printed sensor, and a Wheatstone bridge for temperature sensing was created in [52] using a combination of polymeric materials and a carbon nanoparticle. The printing procedure was carried out at room temperature and with a 2 kHz jetting frequency. The researchers stated that measuring the resistance was a reliable way to assess the sensor's dependability, and they observed high linearity and minimal hysteresis. The production of an inkjet-printed temperature sensor is described in [53].

Stretchable silver conductors were created and positioned on top of a polymeric substrate. After that, the sensors were exposed to ambient and inert gas, and a function of their resistance to temperature was discovered. The studies revealed that the environment had a significant impact on the outcomes, thus coating was utilised to safeguard the sensor and lessen the impact of the ambient temperature.

Particle sensors are tools for evaluating pollution because they can identify airborne particulates [54, 55]. The accuracy and working of this type of sensor depends completely on environmental factors [56]. The production of a 3D-printed particle sensor is described in [57]. Based on the VP approach, the researchers designed and created a compact portable particle sensor. They ran tests with various particles, and it was stated that the particles could be detected.

Tactile sensors are used in measuring force and receiving contact information by providing data on contact pressures [58–61]. The additive manufacturing technique has been used in the flexible manufacturing of the sensors [62, 63]. The sensing components were inserted into an elastic framework as the sensor was constructed layer by layer. The elastomeric material employed in the sensor's construction and sensing layers boosted the sensor's mechanical toughness and flexibility. Carbon nanotube sensors fabricated via a 3D printing method proved to have higher electrical conductivity [64]. The sensitivity was further improved by using two different composites. Subsequently, PLA and nanocomposite materials were used in the production of small sized sensors using FDM and DIW, respectively [65, 66].

6.3.2 Biomedical sensors

Enzyme sensors and DNA sensors are the two primary categories for biomolecule-based sensors. The authors of [67] describe the creation of a 3D-printed lactate sensor with a chemiluminescence sensor for enzymatic assessment that could measure lactate levels in sweat and oral fluid. Microfluidic devices were created using 3D printing to detect glucose levels [68]. The technology could track variations in glucose levels. A microfluidic device was created using 3D printing in [69] to measure the levels of lactate and glucose at the same time. DNA sensors were made using 3D printing methods. In this situation, DNA-coated graphite electrodes were created using the VP approach [70]. A sensitive system for measuring lactate and glucose was built using the ADM approach. The researchers built bioreactors using the FDM method and ABS material [71]. Although there are several ways to make microfluidic devices, 3D printing technology may be employed as a one-step procedure. The ability of 3D printing technology to create arbitrary specified structures using data from CAD software can significantly speed up the creation of microfluidic devices.

Microorganisms that are suitable for detecting organic chemicals and an electro-chemical device have been used to create microbial sensors [72]. Engineering researchers have suggested several approaches depending on the requirements, such as designing and creating microorganism biosensors [73–75]. An investigation into the design of 3D-printed microfluidic chips was utilised to comprehend the gliding motion of cyanobacteria. In this regard, the researchers created variously shaped, embedded-in-glass 3D-printed microfluidics [76]. A microfluidic chip that was 3D-printed was created and employed for influenza virus identification [77]. A 3D microfluidic device was designed for the purpose of finding bacterial pathogens. The components were created using the VP technique and UV light to solidify them. The functional components of the printed microfluidic system were as follows: (i) pillars to filter out large particles, (ii) a particle collector to gather bacteria, and (iii) a chamber for the preconcentration of bacteria.

Living cells provide signals that contain important data about their cell biology and function. This information is crucial for tailored medication and medical diagnostics [78]. Living cells have been included into 3D-printed cell-based sensors in this context, and these sensors may be used for a variety of biological detections. A 3D-printed micro-dissector was utilised to observe cellular ageing at the microscopic level and fluorescence detecting system-based microfluidic devices and micron-scale features were created [79]. It was reported in [80] that a unique cell-based sensor was produced using 3D printing technology. In brief, the researchers created a sensor with gold electrodes and a chip holder using 3D printing. ABS material was processed using the VP method to create the holder, which offers controlled movement. The manufactured sensor was used for the quick detection of alkaline phosphatase from colon cancer cell lines. A 3D cell-based sensor printed using the PBF approach was also created and tested at the same time and is described in [81]. In short, the researchers created a cell-density sensor using polyamide in the PBF process. Valves and pumps were made using the VP method, and a resin-based photopolymer was utilised to create a light-addressable potentiometric sensor in [82]. A 3D-printed cell toxicity sensor's creation and design procedure are described in [83]. The researchers created a smartphone adapter and cell minicartridges using ABS material and the FDM method.

The use of mechanical and electronic components to assist people at various phases of life is referred to as 'bionics' [84]. By fusing novel materials, electronics, and biological systems, the distinctive bionic gadgets may be created. Recent years have shown the capabilities of 3D printing technology in this context [85–87]. The production of a bionic sensor using 3D printing was carried out in [88]. Details of a printed bionic sensor that was utilised in psychological testing and was matched with the epicardium of the heart are provided in [89]. The researchers created an elastic membrane with integrated optoelectronic and sensor components using the MJ process.

6.4 Composite sensors

Composites with improved performance have been created using 3D printing materials, which by nature have low mechanical strength and durability [90–92]. Industries such the automotive and aerospace industries have used these types of materials widely. Improved mechanical properties were achieved by adding carbon fibres into thermoplastic resin by Ning *et al* [90]. By manipulating the orientation of the layer during the printing process, 3D printing of composites may make it possible to digitally include physical characteristics such as composition, stiffness, and toughness into design components. It has also been documented how the incorporation of carbon black into 3D FDM printed resins led to the creation of electrically conductive composite elastomers [21].

Numerous functioning electronic sensors, including capacitive and piezoresistive ones that can detect the presence and amount of liquid, as well as piezoresistive ones that can detect mechanical flexing, have been created. The creation of flex sensors and gloves that can sense mechanical stress served as a demonstration of the material's utility [21]. Adding barium titanate nanoparticles into polyethylene glycol diacrylate solutions led to the development of a piezoelectric nanoparticle polymer composite material which can further be developed into printable electronic sensors [93].

3D-printed sensors have been fabricated in polymer structures to record forces, vibrations, and displacements via composite filaments or fibre wires or other components [37, 94–96]. By aerosol-printing strain sensors with a silver nanoparticle ink directly onto carbon fibre prepreg, Zhao et al [97] created multifunctional composites with sensing capabilities. Leigh et al developed a composite of carbon black filler with a polycaprolactone matrix which was used to 3D print electronic sensors [21]. These sensors were fabricated via the FDM process to sense mechanical flexing and capacitance variations. Other sensors such as flow sensors were printed using FDM which showed a more satisfying result in comparison to the performance of a commercially available sensor [98]. Using methods such as FEAM and TEAM, a composite capacitive force sensor was made to detect applied forces [44].

Using acrylonitrile butadiene styrene and clear polylactic acid as the filament in 3D printing of high-tech electronic sensors has the advantages of being cost efficient and producing less complex circuits that can be fabricated on a desktop printer without the need for production areas [21]. Kennedy et al worked on 3D-printed polyvinylidene fluoride (PVDF) and carbon nanotube composites in the production of chemical vapour sensors [99]. Changes in resistance during exposure were used to track how quickly (20 s) and irreversibly (through at least 25 cycles) the printed components were still able to detect volatile organic compound (VOC) vapours. Xiang et al studied a carbon nanotube and graphene nanoplatelet thermoplastic strain sensor fabricated via fused filament 3D printing [100]. There has been an increase in the popularity of stretchable temperature sensors due to their wide range of applications in biomedical devices, electronic devices, food monitoring, and many others [101–106]. Graphene polydimethysiloxane composites have been used in the production of stretchable temperature sensors by Wang et al [107]. The product proved to have excellent performance in measuring the temperature of surfaces, such as heated tube or skin surfaces, providing accurate results. To create uniaxial and biaxial strain sensors with integrated conductive routes in the insulative TPU substrate, pure thermoplastic polyurethane (TPU) and TPU/multiwall carbon nanotube (MWCNT) nanocomposites were 3D-printed simultaneously [94]. Following that, a series of cyclic strain loads were applied to the sensors. The sensors' outstanding piezoresistive responses to stresses as high as 50% were shown by the findings, which also showed cyclic repeatability in both the axial and transverse directions.

Table 6.1. Overview of different materials for the 3D printing process.

Additive manufacturing technology	Sensor type	Material used	Application
Fused deposition modelling (FDM)	Strain gauge	Thermoplastic	Structural health monitoring
	Pressure sensor	Thermoplastic	Healthcare monitoring
	Chemical sensor	Conductive polymer	Environmental monitoring
Direct ink writing (DIW)	Stretchable sensor	Silver nanoparticle ink	Wearable electronics
	Temperature sensor	Carbon nanotube ink	Aerospace
Stereolithography (SLA)	Gas sensor	Polydimethylsiloxane (PDMS)	Environmental monitoring
	Biosensor	Hydrogel	Medical diagnostics
Selective laser sintering (SLS)	Gas sensor	Metal oxide powder	Environmental monitoring
	Acoustic sensor	Thermoplastic	Structural health monitoring
Inkjet printing	Gas sensor	Zinc oxide nanoparticles	Environmental monitoring
	Humidity sensor	Silver nanoparticles	HVAC systems
Binder jetting	pH sensor	Glass-ceramic	Chemical processing
Material jetting	Biosensor	Photopolymer	Medical diagnostics
Digital light processing (DLP)	Temperature sensor	Epoxy resin	Aerospace
Laser powder bed fusion (LPBF)	Pressure sensor	Stainless steel	Industrial process monitoring
Fused filament fabrication (FFF)	Biosensor	Chitosan	Medical diagnostics
	Gas sensor	Carbon black/ Polycarbonate	Environmental monitoring

6.5 Materials used in the AM process

Each additive manufacturing process has its own properties and constraints in utilizing materials. An overview of different materials in the forms of wires, filaments, powder, and paste is shown in table 6.1 for various 3D printing process in engineering and biomedical applications [108–125].

6.6 Cost effectiveness

The cost effectiveness of additive manufacturing compared to other manufacturing processes depends on several factors, such as the type of part being produced, the quantity of parts needed, the material used, and the required level of precision and surface finish. Here are some general considerations when comparing the cost effectiveness of additive manufacturing with traditional manufacturing processes:

1. **Tooling costs:** Traditional manufacturing processes often require the use of specialized tooling, moulds, and fixtures, which can be expensive to produce. Additive manufacturing, on the other hand, does not require tooling, which can be cost-effective for small batch runs or customized parts.

2. **Material waste:** Traditional manufacturing processes often produce more material waste than additive manufacturing. This can increase the cost of raw materials and the cost of disposing of waste materials.

3. **Labour costs:** Traditional manufacturing processes often require more labour than additive manufacturing, which can increase costs. However, the cost of skilled labour for additive manufacturing may be higher than for some traditional processes.

4. **Part complexity:** Additive manufacturing is often more cost-effective for parts with complex geometries or designs that are difficult to produce using traditional manufacturing processes.

6. **Production volume:** Traditional manufacturing processes may be more cost-effective for high-volume production runs, while additive manufacturing may be more cost-effective for low-volume production runs or customized parts.

3D printing technology provides low cost and precise sensors that are being used for various purposes [126–128]. In [129], polyurethane thermoplastic was used for the production of 3D-printed strain sensors and different properties, such as electrical, mechanical, and piezoresistive, were tested. Based on the results obtained, AM was chosen as the most suitable choice where customizing and complexity are involved. Based on the study [130], on a displacement sensor produced using the inkjet printing technique, it was found that the precision, design, and low cost of fabrication were advantages in the printing process. The cost model of Ruffo, Tuck, and Hague shows how the cost of a 3D-printed material is calculated [131].

Overall, the cost effectiveness of additive manufacturing versus traditional manufacturing processes will depend on the specific application and requirements of the part being produced. In some cases, additive manufacturing may be more cost-effective, while in others, traditional manufacturing processes may be more cost-effective.

6.7 Challenges and future prospects

The market for printed electronics might reach $52.82 billion by 2031, according to IDTechEx research [132]. This shows how 3D printing might be used, among other things, in sensors and electronic equipment. Electronic device design and printing technology must be coupled to create a 3D-printed sensor. Mechanical, electrical, and material science competence and understanding are required for this combination.

In this quick prototyping process, material limitations in each 3D printing method might be seen as a hurdle. In the previous section, we outlined the materials that are currently available for each AM technique. The reliability and performance of printed sensors are affected by this material constraint. In order to enhance the production of 3D-printed sensors, the limitations in currently available printable materials need to be addressed.

The life cycle of these 3D-printed electronic components still needs to be examined due to a lack of information on the durability of these products. The amount of waste from the electronic industries is huge, so 3D-printed sensors with shorter lifespans could lead to more waste being generated.

According to studies and reports there are only a few selected composite materials that can be printed into electronic sensors. While the basic specifications are easily fulfilled using such composites, some disadvantages still remain, such as the high cost and temperature dependency. There is a lot of research potential in this area of material development for manufacturing different sensors with high efficiency. Technical research and development are still needed to use 3D printing to fabricate sensors. Future uses of 3D-printed sensors are anticipated due to the growing needs in the engineering and medical fields.

Future research and development might focus on topics including the creation of biodegradable materials for sensors, enhancing 3D-printed component adhesive connections, and developing new printing techniques with greater resolutions. Furthermore, as 3D-printed sensors employ nanomaterials, future research will need to address the issue of recycling nano-waste. Studying the impact of 3D-printed sensors' physical, chemical, and geometric properties on their performance is advised. This research can be done independently or in combination. Further analysis and comparison of using 3D printing techniques to create the same sensor are currently lacking.

6.8 Conclusion

With the advent of 3D printing technology, cheap, sensitive, small sensors can be fabricated easily, as can be seen from the works discussed. Here the use of 3D printing for the manufacture of electronic sensors is discussed with composites being a prime material for 3D-printed sensors. We discuss some academic and commercial studies that made use of 3D printing to create various sensors. There are some recommendations provided along with problems and potential outcomes. The wide range of 3D-printed sensors, from polymer-based to metals to composites, addresses the needs of engineering and medicine. The 3D-printed sensors may be made more responsive, flexible, and sensitive by streamlining the production process.

References

[1] Agarwala S, Goh G L, Yap Y L, Goh G D, Yu H, Yeong W Y and Tran T 2017 Development of bendable strain sensor with embedded microchannels using 3D printing *Sensors Actuators* A **263** 593–9

[2] Oliveira J, Correia V, Castro H, Martins P and Lanceros-Mendez S 2018 Polymer-based smart materials by printing technologies: improving application and integration *Addit. Manuf.* **21** 269–83

[3] Liu C, Huang N, Xu F, Tong J, Chen Z, Gui X, Fu Y and Lao C 2018 3D printing technologies for flexible tactile sensors toward wearable electronics and electronic skin *Polymers* **10** 629

[4] Wang Y, Wang Y, Liu W, Chen D, Wu C and Xie J 2019 An aerosol sensor for PM1 concentration detection based on 3D printed virtual impactor and SAW sensor *Sensors Actuators* A **288** 67–74

[5] Zolfagharian A, Khosravani M R and Kaynak A 2020 Fracture resistance analysis of 3D-printed polymers *Polymers* **12** 302

[6] Ivanova O, Williams C and Campbell T 2013 Additive manufacturing (AM) and nanotechnology: promises and challenges *Rapid Prototyp. J.* **19** 353–64

[7] Upcraft S and Fletcher R 2003 The rapid prototyping technologies *Assem. Autom.* **23** 318–30

[8] Williams J M, Adewunmi A, Schek R M, Flanagan C L, Krebsbach P H, Feinberg S E, Hollister S J and Das S 2005 Bone tissue engineering using polycaprolactone scaffolds fabricated via selective laser sintering *Biomaterials* **26** 4817–27

[9] Li X, Wang C, Zhang W and Li Y 2009 Fabrication and characterization of porous Ti6Al4V parts for biomedical applications using electron beam melting process *Mater. Lett.* **63** 403–5

[10] Zhang X, Jiang X N and Sun C 1999 Micro-stereolithography of polymeric and ceramic microstructures *Sensors Actuators* A **77** 149–56

[11] Zein I, Hutmacher D W, Tan K C and Teoh S H 2002 Fused deposition modeling of novel scaffold architectures for tissue engineering applications *Biomaterials* **23** 1169–85

[12] Jahangir M N, Cleeman J, Hwang H J and Malhotra R 2019 Towards out-of-chamber damage-free fabrication of highly conductive nanoparticle-based circuits inside 3D printed thermally sensitive polymers *Addit. Manuf.* **30** 100886

[13] Zhang F, Saleh E, Vaithilingam J, Li Y, Tuck C J, Hague R J, Wildman R D and He Y 2019 Reactive material jetting of polyimide insulators for complex circuit board design *Addit. Manuf.* **25** 477–84

[14] Zolfagharian A, Kaynak A and Kouzani A 2020 Closed-loop 4D-printed soft robots *Mater. Des.* **188** 108411

[15] Lipomi D J, Vosgueritchian M, Tee B C, Hellstrom S L, Lee J A, Fox C H and Bao Z 2011 Skin-like pressure and strain sensors based on transparent elastic films of carbon nanotubes *Nat. Nanotechnol.* **6** 788–92

[16] Aremu A O, Brennan-Craddock J P, Panesar A, Ashcroft I A, Hague R J, Wildman R D and Tuck C 2017 A voxel-based method of constructing and skinning conformal and functionally graded lattice structures suitable for additive manufacturing *Addit. Manuf.* **13** 1–3

[17] Tofail S A, Koumoulos E P, Bandyopadhyay A, Bose S, O'Donoghue L and Charitidis C 2018 Additive manufacturing: scientific and technological challenges, market uptake and opportunities *Mater. Today* **21** 22–37

[18] O'Neill P F, Ben Azouz A, Vazquez M, Liu J, Marczak S, Slouka Z, Chang H C, Diamond D and Brabazon D 2014 Advances in three-dimensional rapid prototyping of microfluidic devices for biological applications *Biomicrofluidics* **8** 052112

[19] Christ J F, Aliheidari N, Ameli A and Pötschke P 2017 3D printed highly elastic strain sensors of multiwalled carbon nanotube/thermoplastic polyurethane nanocomposites *Mater. Des.* **131** 394–401

[20] Lu J R, Weng W G, Chen X F, Wu D J, Wu C L and Chen G H 2005 Piezoresistive materials from directed shear-induced assembly of graphite nanosheets in polyethylene *Adv. Funct. Mater.* **15** 1358–63

[21] Leigh S J, Bradley R J, Pursell C P, Billson D R and Hutchins D A 2012 A simple, low-cost conductive composite material for 3D printing of electronic sensors *PLoS One* **7** e49365

[22] Khosravani M R and Reinicke T 2020 On the environmental impacts of 3D printing technology *Appl. Mater. Today* **20** 100689

[23] ASTM Committee F42 on Additive Manufacturing Technologies 2012 *Standard Terminology for Additive Manufacturing Technologies* ASTM International

[24] Sirringhaus H, Kawase T, Friend R H, Shimoda T, Inbasekaran M, Wu W and Woo E P 2000 High-resolution inkjet printing of all-polymer *Transistor Circuits* **29** 5

[25] Maurizi M, Cianetti F, Slavič J, Zucca G and Palmieri M 2019 Piezoresistive dynamic simulations of FDM 3D-printed embedded strain sensors: a new modal approach *Procedia Struct. Integrity* **24** 390–7

[26] Shamsinejad S, De Flaviis F and Mousavi P 2015 Microstrip-fed 3-D folded slot antenna on cubic structure *IEEE Antennas Wirel. Propag. Lett.* **15** 1081–4

[27] Wan X, Zhang F, Liu Y and Leng J 2019 CNT-based electro-responsive shape memory functionalized 3D printed nanocomposites for liquid sensors *Carbon* **155** 77–87

[28] Pang C, Lee C and Suh K Y 2013 Recent advances in flexible sensors for wearable and implantable devices *J. Appl. Polym. Sci.* **130** 1429–41

[29] Jibril L, Ramírez J, Zaretski A V and Lipomi D J 2017 Single-nanowire strain sensors fabricated by nanoskiving *Sensors Actuators* A **263** 702–6

[30] Larimi S R, Nejad H R, Oyatsi M, O'Brien A, Hoorfar M and Najjaran H 2018 Low-cost ultra-stretchable strain sensors for monitoring human motion and bio-signals *Sensors Actuators* A **271** 182–91

[31] Khosravani M R, Anders D and Weinberg K 2019 Influence of strain rate on fracture behavior of sandwich composite T-joints *Eur. J. Mech.* A **78** 103821

[32] Mattmann C, Clemens F and Tröster G 2008 Sensor for measuring strain in textile *Sensors* **8** 3719–32

[33] Herrmann J, Müller K H, Reda T, Baxter G R, Raguse B D, De Groot G J, Chai R, Roberts M and Wieczorek L 2007 Nanoparticle films as sensitive strain gauges *Appl. Phys. Lett.* **91** 183105

[34] Yamada T, Hayamizu Y, Yamamoto Y, Yomogida Y, Izadi-Najafabadi A, Futaba D N and Hata K 2011 A stretchable carbon nanotube strain sensor for human-motion detection *Nat. Nanotechnol.* **6** 296–301

[35] Zhao J, Zhang G Y and Shi D X 2013 Review of graphene-based strain sensors *Chin. Phys.* B **22** 057701

[36] Giffney T, Bejanin E, Kurian A S, Travas-Sejdic J and Aw K 2017 Highly stretchable printed strain sensors using multi-walled carbon nanotube/silicone rubber composites *Sensors Actuators* A **259** 44–9

[37] Muth J T, Vogt D M, Truby R L, Mengüç Y, Kolesky D B, Wood R J and Lewis J A 2014 Embedded 3D printing of strain sensors within highly stretchable elastomers *Adv. Mater.* **26** 6307–12

[38] Bodnicki M, Pakuła P and Zowade M 2016 Miniature displacement sensor *Advanced Mechatronics Solutions* (Cham: Springer) pp 313–8

[39] Macdonald E, Salas R, Espalin D, Perez M, Aguilera E, Muse D and Wicker R B 2014 3D printing for the rapid prototyping of structural electronics *IEEE Access* **2** 234–42

[40] Cappelleri D J, Piazza G and Kumar V 2011 A two dimensional vision-based force sensor for microrobotic applications *Sensors Actuators* A **171** 340–51

[41] Roriz P, Carvalho L, Frazão O, Santos J L and Simões J A 2014 From conventional sensors to fibre optic sensors for strain and force measurements in biomechanics applications: a review *J. Biomech.* **47** 1251–61

[42] Buttafuoco A, Lenders C, Clavel R, Lambert P and Kinnaert M 2014 Design, manufacturing and implementation of a novel 2-axis force sensor for haptic applications *Sensors Actuators* A **209** 107–14

[43] Zang H, Zhang X, Zhu B and Fatikow S 2019 Recent advances in non-contact force sensors used for micro/nano manipulation *Sensors Actuators* A **296** 155–77

[44] Saari M, Xia B, Cox B, Krueger P S, Cohen A L and Richer E 2016 Fabrication and analysis of a composite 3D printed capacitive force sensor *3D Print. Addit. Manuf.* **3** 136–41

[45] Kesner S B and Howe R D 2011 Design principles for rapid prototyping forces sensors using 3-D printing *IEEE/ASME Trans. Mechatron.* **16** 866–70

[46] Shi H L 2015 A narrow vertical beam based structure for passive pressure measurement using two-material 3D printing *Proc. of the World Congress on Engineering and Computer Science* vol 1

[47] Gao Y, Xu M, Yu G, Tan J and Xuan F 2019 Extrusion printing of carbon nanotube-coated elastomer fiber with microstructures for flexible pressure sensors *Sensors Actuators* A **299** 111625

[48] Emon M O, Alkadi F, Philip D G, Kim D H, Lee K C and Choi J W 2019 Multi-material 3D printing of a soft pressure sensor *Addit. Manuf.* **28** 629–38

[49] Courbat J, Kim Y B, Briand D and De Rooij N F 2011 Inkjet printing on paper for the realization of humidity and temperature sensors *16th International Solid-State Sensors, Actuators and Microsystems Conf.* (Piscataway, NJ: IEEE) pp 1356–9

[50] Wickberg A, Mueller J B, Mange Y J, Fischer J, Nann T and Wegener M 2015 Three-dimensional micro-printing of temperature sensors based on up-conversion luminescence *Appl. Phys. Lett.* **106** 133103

[51] Dankoco M D, Tesfay G Y, Benevent E and Bendahan M 2016 Temperature sensor realized by inkjet printing process on flexible substrate *Mater. Sci. Eng.* B **205** 1–5

[52] Bali C, Brandlmaier A, Ganster A, Raab O, Zapf J and Hübler A 2016 Fully inkjet-printed flexible temperature sensors based on carbon and PEDOT:PSS *Mater. Today Proc.* **3** 739–45

[53] Vuorinen T, Niittynen J, Kankkunen T, Kraft T M and Mäntysalo M 2016 Inkjet-printed graphene/PEDOT:PSS temperature sensors on a skin-conformable polyurethane substrate *Sci. Rep.* **6** 1–8

[54] Tang D, Zhao R, Wang S, Wang J, Ni L and Chen L 2017 The simulation and experimental research of particulate matter sensor on diesel engine with diesel particulate filter *Sensors Actuators* A **259** 160–70

[55] Zusman M, Schumacher C S, Gassett A J, Spalt E W, Austin E, Larson T V, Carvlin G, Seto E, Kaufman J D and Sheppard L 2020 Calibration of low-cost particulate matter sensors: model development for a multi-city epidemiological study *Environ. Int.* **134** 105329

[56] Wang Y, Li J, Jing H, Zhang Q, Jiang J and Biswas P 2015 Laboratory evaluation and calibration of three low-cost particle sensors for particulate matter measurement *Aerosol Sci. Technol.* **49** 1063–77

[57] Wang Y, Mackenzie F V, Ingenhut B and Boersma A 2018 AP4.1-miniaturized 3D printed particulate matter sensor for personal monitoring *Proceedings IMCS 2018* pp 402–3

[58] Hasegawa Y, Shikida M, Shimizu T, Miyaji T, Sasaki H, Sato K and Itoigawa K 2004 A micromachined active tactile sensor for hardness detection *Sensors Actuators* A **114** 141–6

[59] Tiwana M I, Redmond S J and Lovell N H 2012 A review of tactile sensing technologies with applications in biomedical engineering *Sensors Actuators* A **179** 17–31

[60] Sohgawa M, Hirashima D, Moriguchi Y, Uematsu T, Mito W, Kanashima T, Okuyama M and Noma H 2012 Tactile sensor array using microcantilever with nickel–chromium alloy thin film of low temperature coefficient of resistance and its application to slippage detection *Sensors Actuators* A **186** 32–7

[61] Girão P S, Ramos P M, Postolache O and Pereira J M 2013 Tactile sensors for robotic applications *Measurement* **46** 1257–71

[62] Vatani M, Engeberg E D and Choi J W 2013 Force and slip detection with direct-write compliant tactile sensors using multi-walled carbon nanotube/polymer composites *Sensors Actuators* A **195** 90–7

[63] Vatani M, Engeberg E D and Choi J W 2015 Conformal direct-print of piezoresistive polymer/nanocomposites for compliant multi-layer tactile sensors *Addit. Manuf.* **7** 73–82

[64] Yun H Y, Kim H C and Lee I H 2015 Research for improved flexible tactile sensor sensitivity *J. Mech. Sci. Technol.* **29** 5133–8

[65] Ota H *et al* 2016 Application of 3D printing for smart objects with embedded electronic sensors and systems *Adv. Mater. Technol.* **1** 1600013

[66] Shi G, Lowe S E, Teo A J, Dinh T K, Tan S H, Qin J, Zhang Y, Zhong Y L and Zhao H 2019 A versatile PDMS submicrobead/graphene oxide nanocomposite ink for the direct ink writing of wearable micron-scale tactile sensors *Appl. Mater. Today* **16** 482–92

[67] Roda A, Guardigli M, Calabria D, Calabretta M M, Cevenini L and Michelini E 2014 A 3D-printed device for a smartphone-based chemiluminescence biosensor for lactate in oral fluid and sweat *Analyst* **139** 6494–501

[68] Comina G, Suska A and Filippini D 2015 Autonomous chemical sensing interface for universal cell phone readout *Angew. Chem.* **127** 8832–6

[69] Gowers S A, Curto V F, Seneci C A, Wang C, Anastasova S, Vadgama P, Yang G Z and Boutelle M G 2015 3D printed microfluidic device with integrated biosensors for online analysis of subcutaneous human microdialysate *Anal. Chem.* **87** 7763–70

[70] Bishop G W, Satterwhite-Warden J E, Bist I, Chen E and Rusling J F 2016 Electrochemiluminescence at bare and DNA-coated graphite electrodes in 3D-printed fluidic devices *ACS Sens.* **1** 197–202

[71] Loo A H, Chua C K and Pumera M 2017 DNA biosensing with 3D printing technology *Analyst* **142** 279–83

[72] Su C K, Yen S C, Li T W and Sun Y C 2016 Enzyme-immobilized 3D-printed reactors for online monitoring of rat brain extracellular glucose and lactate *Anal. Chem.* **88** 6265–73

[73] Reyes D R, Iossifidis D, Auroux P A and Manz A 2002 Micro total analysis systems. 1. Introduction, theory, and technology *Anal. Chem.* **74** 2623–36

[74] Lim J J, Malheiros L R, Bertali G, Long C J, Freyer P D and Burke M G 2015 Comparison of additive manufactured and conventional 316L stainless steels–ERRATUM *Microsc. Microanal.* **16** 65

[75] Liu Y, Zhou H, Hu Z, Yu G, Yang D and Zhao J 2017 Label and label-free based surface-enhanced Raman scattering for pathogen bacteria detection: a review *Biosens. Bioelectron.* **94** 131–40

[76] Hanada Y, Sugioka K, Shihira-Ishikawa I, Kawano H, Miyawaki A and Midorikawa K 2011 3D microfluidic chips with integrated functional microelements fabricated by a femtosecond laser for studying the gliding mechanism of cyanobacteria *Lab Chip* **11** 2109–15

[77] Krejcova L, Nejdl L, Rodrigo M A, Zurek M, Matousek M, Hynek D, Zitka O, Kopel P, Adam V and Kizek R 2014 3D printed chip for electrochemical detection of influenza virus labeled with CdS quantum dots *Biosens. Bioelectron.* **54** 421–7

[78] Ragones H, Schreiber D, Inberg A, Berkh O, Kósa G and Shacham-Diamand Y 2015 Processing issues and the characterization of soft electrochemical 3D sensor *Electrochim. Acta* **183** 125–9

[79] Spivey E C, Xhemalce B, Shear J B and Finkelstein I J 2014 3D-printed microfluidic microdissector for high-throughput studies of cellular aging *Anal. Chem.* **86** 7406–12

[80] Ragones H, Schreiber D, Inberg A, Berkh O, Kósa G, Freeman A and Shacham-Diamand Y 2015 Disposable electrochemical sensor prepared using 3D printing for cell and tissue diagnostics *Sensors Actuators* B **216** 434–42

[81] Ude C, Hentrop T, Lindner P, Lücking T H, Scheper T and Beutel S 2015 New perspectives in shake flask pH control using a 3D-printed control unit based on pH online measurement *Sensors Actuators* B **221** 1035–43

[82] Takenaga S, Schneider B, Erbay E, Biselli M, Schnitzler T, Schöning M J and Wagner T 2015 Fabrication of biocompatible lab-on-chip devices for biomedical applications by means of a 3D-printing process *Physica Status Solidi* a **212** 1347–52

[83] Cevenini L, Calabretta M M, Tarantino G, Michelini E and Roda A 2016 Smartphone-interfaced 3D printed toxicity biosensor integrating bioluminescent 'sentinel cells *Sensors Actuators* B **225** 249–57

[84] Kong Y L, Gupta M K, Johnson B N and McAlpine M C 2016 3D printed bionic nanodevices *Nano Today* **11** 330–50

[85] Gross B C, Erkal J L, Lockwood S Y, Chen C and Spence D M 2014 Evaluation of 3D printing and its potential impact on biotechnology and the chemical sciences *Anal. Chem.* **86** 3240–53

[86] Johnson B N, Lancaster K Z, Hogue I B, Meng F, Kong Y L, Enquist L W and McAlpine M C 2016 3D printed nervous system on a chip *Lab Chip* **16** 1393

[87] Zhang L, Yang G, Johnson B N and Jia X 2019 Three-dimensional (3D) printed scaffold and material selection for bone repair *Acta Biomater.* **84** 16–33

[88] Mannoor M S, Jiang Z, James T, Kong Y L, Malatesta K A, Soboyejo W O, Verma N, Gracias D H and McAlpine M C 2013 3D printed bionic ears *Nano Lett.* **13** 2634–9

[89] Xu L, Zhang L, Du L and Zhang S 2014 Electro-catalytic oxidation in treating CI Acid Red 73 wastewater coupled with nanofiltration and energy consumption analysis. *J. Membr. Sci.* **452** 1–0

[90] Ning F, Cong W, Qiu J, Wei J and Wang S 2015 Additive manufacturing of carbon fiber reinforced thermoplastic composites using fused deposition modeling *Composites* **B 80** 369–78

[91] Compton B G and Lewis J A 2014 3D-printing of lightweight cellular composites *Adv. Mater.* **26** 5930–5

[92] Chuang K C, Grady J E, Draper R D, Shin E S, Patterson C and Santelle T D 2015 Additive manufacturing and characterization of Ultem polymers and composites *The Composites and Advanced Materials Expo CAMX (Dallas, TX, 26 October 2016)* 20160001352

[93] Kim K, Zhu W, Qu X, Aaronson C, McCall W R, Chen S and Sirbuly D J 2014 3D optical printing of piezoelectric nanoparticle–polymer composite materials *ACS Nano* **8** 9799–806

[94] Christ J F, Hohimer C J, Aliheidari N, Ameli A, Mo C and Pötschke P 2017 3D printing of highly elastic strain sensors using polyurethane/multiwall carbon nanotube composites *Proc. SPIE* **10168** 101680E

[95] Shemelya C, Cedillos F, Aguilera E, Espalin D, Muse D, Wicker R and MacDonald E 2014 Encapsulated copper wire and copper mesh capacitive sensing for 3-D printing applications *IEEE Sens. J.* **15** 1280–6

[96] Shemelya C, Banuelos-Chacon L, Melendez A, Kief C, Espalin D, Wicker R, Krijnen G and MacDonald E 2015 Multi-functional 3D printed and embedded sensors for satellite qualification structures *2015 IEEE Sensors* (Piscataway, NJ: IEEE) pp 1–4

[97] Zhao D, Liu T, Zhang M, Liang R and Wang B 2012 Fabrication and characterization of aerosol-jet printed strain sensors for multifunctional composite structures *Smart Mater. Struct.* **21** 115008

[98] Leigh S J, Purssell C P, Billson D R and Hutchins D A 2014 Using a magnetite/thermoplastic composite in 3D printing of direct replacements for commercially available flow sensors *Smart Mater. Struct.* **23** 095039

[99] Kennedy Z C, Christ J F, Evans K A, Arey B W, Sweet L E, Warner M G, Erikson R L and Barrett C A 2017 3D-printed poly (vinylidene fluoride)/carbon nanotube composites as a tunable, low-cost chemical vapour sensing platform *Nanoscale* **9** 5458–66

[100] Xiang D *et al* 2020 3D printed high-performance flexible strain sensors based on carbon nanotube and graphene nanoplatelet filled polymer composites *J. Mater. Sci.* **55** 15769–86

[101] Trung T Q and Lee N E 2016 Flexible and stretchable physical sensor integrated platforms for wearable human-activity monitoring and personal healthcare *Adv. Mater.* **28** 4338–72

[102] Liu Y, Pharr M and Salvatore G A 2017 Lab-on-skin: a review of flexible and stretchable electronics for wearable health monitoring *ACS Nano* **11** 9614–35

[103] Salvatore G A *et al* 2017 Biodegradable and highly deformable temperature sensors for the Internet of things *Adv. Funct. Mater.* **27** 1702390

[104] Guo H, Yeh M H, Zi Y, Wen Z, Chen J, Liu G, Hu C and Wang Z L 2017 Ultralight cut-paper-based self-charging power unit for self-powered portable electronic and medical systems *ACS Nano* **11** 4475–82

[105] Wang X and Yang Y 2017 Effective energy storage from a hybridized electromagnetic-triboelectric nanogenerator *Nano Energy* **32** 36–41

[106] Sima K, Syrovy T, Pretl S, Freisleben J, Cesek D and Hamacek A 2017 Flexible smart tag for cold chain temperature monitoring *2017 40th Int. Spring Seminar on Electronics Technology (ISSE)* (Piscataway, NJ: IEEE) pp 1–5

[107] Wang Z, Gao W, Zhang Q, Zheng K, Xu J, Xu W, Shang E, Jiang J, Zhang J and Liu Y 2018 3D-printed graphene/polydimethylsiloxane composites for stretchable and strain-insensitive temperature sensors *ACS Appl. Mater. Interfaces* **11** 1344–52

[108] Park Y L, Chen B R and Wood R J 2012 Design and fabrication of soft artificial skin using embedded microchannels and liquid conductors *IEEE Sens. J.* **12** 2711–8

[109] Xiang D, Zhang X, Hakin-Jones E, Zhu W, Zhou Z, Shen Y, Li Y, Zhao C and Wang P 2020 Synergistic effects of hybrid conductive nanofillers on the performance of 3D printed highly elastic strain sensors *Composites* A **129** 105730

[110] Kim K *et al* 2017 3D printing of multi axial force sensors using carbon nanotube (CNT)/thermoplastic polyurethane (TPU) filaments *Sensors Actuators* A **263** 493–500

[111] Wassefall F, Hendrich N, Fiedler F and Zhang J 2017 3D-printed low-cost modular force sensors *Proc. of Int. Conf. on Climbing and Walking Robots and the Support Technologies for Mobile Machines (Porto, Portugal)* pp 1–8

[112] Zega V, Credi C, Bernasconi R, Langfelder G, Magagnin L, Levi M and Corigliano A 2018 The first 3-D-printed z-axis accelerometers with differential capacitive sensing *IEEE Sens. J.* **18** 53–60

[113] Zega V, Invernizzi M, Bernasconi R, Cuneo F, Langfelder G, Magagnin L, Levi M and Corigliano A 2019 The first 3D-printed and wet-metallized three-axis accelerometer with differential capacitive sensing *IEEE Sens. J.* **19** 9131–8

[114] Suaste Gomez E, Roldan G R, Reyes Curz H and Jimenez O T 2016 Developing an ear prosthesis fabricated in polyvinylidene fluoride by a 3D printer with sensory intrinsic properties of pressure and temperature *Sensors* **16** 332

[115] Lin Y, Hsieh T, Tsai L, Wang S and Chiang C 2016 Using three-dimensional printing technology to produce a novel optical fiber Bragg grating pressure sensor *Sens. Mater.* **28** 389–94

[116] Sauerbrunn E, Chen Y, Didion J, Yu M, Smela E and Bruck H A 2015 Thermal imaging using polymer nanocomposite temperature sensors *Phys. Status Solidi* a **212** 2239–45

[117] Wang Z, Gao W, Zhang Q, Zheng K, Xu J, Xu W, Shang E, Jiang J, Zhang J and Liu Y 2019 3D-printed graphene/polydimethylsiloxane composites for stretchable and strain-insensitive temperature sensors *Appl. Mater. Interfaces.* **11** 1344–52

[118] Marzo A M L, Mayorga-Martinez C G and Pumera M 2020 3D printed graphene direct electron transfer enzyme biosensors *Biosens. Bioelectron.* **151** 111980

[119] Damiati S, Küpcü S, Peacock M, Eilenberger C, Zamzami M, Qadri I, Choudhry H, Sleytr U B and Schuster B 2017 Acoustic and hybrid 3D-printed electrochemical biosensors for the realtimeimmunodetection of liver cancer cells (HepG2) *Biosens. Bioelectron.* **94** 500–6

[120] Mandon C A, Blum L J and Marquette C A 2016 Adding biomolecular recognition capability to 3D printed objects *Anal. Chem.* **88** 10767–72

[121] Lind J U *et al* 2017 Instrumented cardiac microphysiological devicesvia multimaterial three-dimensional printing *Nat. Mater.* **16** 303–9

[122] Motaghi H, Ziyaee S, Mehrgardi M A, Kajani A A and Bordbar A 2018 Electrochemiluminescence detection of human breast cancer cells using aptamer modified bipolar electrode mounted into 3D printed microchannel *Biosens. Bioelectron.* **118** 217–23

[123] Dantism S, Takenaga S, Eagner P, Wagner T and Schöning M J 2016 Determination of the extracellular acidification of *Escherichia coli* K12 with a multi-chamber-based LAPS system *Phys. Status Solidi* a **213** 1479–85

[124] Philamore H, Rossiter J, Walters P and Winfeld J 2015 Cast and 3D printed ion exchange membranes for monolithicmicrobial fuel cell fabrication *J. Power Sources* **289** 91–9

[125] Mills D K, Jammalamadaka U, Tappa K and Weisman J 2018 Studies on the cytocompatibility, mechanical and antimicrobialproperties of 3D printed poly (methyl methacrylate) bead *Bioact. Mater.* **3** 157–66

[126] Aremu A O, Brennan-Craddock J P J, Panesar A, Aschcroft I A, Hague R J M, Wildman R D and Tuck C 2017 A voxel-based method of constructing and skinning conformal and functionally graded lattice structures suitable for additive manufacturing *Addit. Manuf.* **13** 1–13

[127] Tofail S A M, Koumoulos E P, Bandyopadhyay A, Bose S, O'Donoghue L and Charitidis C 2018 Additive manufacturing: scientific andtechnological challenges, market uptakeand opportunities *Mater. Today* **21** 22–37

[128] Vilardell A M, Takezawa A, Plessis A d, Takata N, Krakhmalev P, Kobashi M, Yadroitsava I and Yadroitsev I 2019 Topology optimization and characterization of Ti6Al4V ELI cellular lattice structures by laser powder bed fusion for biomedical applications *Sensors* **19** 1–22

[129] Christ J F, Aliheidari N, Ameli A and Pötschke P 2017 3D printed highly elastic strain sensors of multiwalled carbonnanotube/thermoplastic polyurethane nanocomposites *Mater. Des.* **131** 394–401

[130] Jerance N, Bednar N and Stojanovic G 2013 An ink-jet printed eddy current position sensor *Sensors* **13** 5205–19

[131] Ruffo M, Tuck C and Hague R 2006 Cost estimation for rapid manufacturing-laser sintering production for low to medium volumes *Proc. Inst. Mech. Eng.* B **220** 1417–27

[132] Das R and Harrop P 2013 *Printed, Organic and Flexible Electronics: Forecasts, Players and Opportunities 2013–2023* (Cambridge: IDTechEx)

IOP Publishing

3D Printed Smart Sensors and Energy Harvesting Devices
Concepts, fabrication and applications
Sanket Goel and Sohan Dudala

Chapter 7

3D printing for wearable biosensing and energy storage devices

Vijay Vaishampayan, Rakesh Kumar, Ashish Kapoor and Sarang P Gumfekar

Wearable devices with advanced functionalities are emerging as next-generation gadgets for diverse applications. Their desirable characteristics include ultra-portability, light weight, flexibility, cost-effectiveness, user comfort, and integrability with other technologies. In this context, manufacturing using three-dimensional (3D) printing has emerged as a promising technique for developing customized and standalone products. The advantages offered by 3D printing, such as rapid fabrication and the ability to produce precise and reproducible features, give the technique an edge over conventional manufacturing technologies. This chapter aims to provide an overview of recent advancements in 3D-printed wearable devices with a specific focus on biosensing and energy applications. An overview is given of various additive manufacturing techniques. The materials compatible with different techniques are discussed, and their scope and limitations are identified. Illustrative examples are described for biosensors and energy storage devices in terms of design specifications, manufacturing approaches, and user-oriented applications. A critical evaluation is made of the challenges associated with developing 3D-printed wearable devices. Future directions are identified toward the design and development of multifunctional devices integrated with information technologies for widespread user acceptance.

7.1 Introduction

3D printing has drawn significant attention in the last several years due to its versatile properties such as fast processing speed, user-friendliness, and less wastage of material. Various techniques, including the material extrusion method, binder jetting method, photopolymerization method, and sheet lamination approach, are included under 3D printing [1, 2]. These methods offer exceptional flexibility and simplicity in critical designs. The rapid emergence of sensors and energy storage devices following the discovery of several new materials has allowed researchers to create more innovative products. Developing microelectronic sensors on flexible substrates could become a cost-effective and creative way of monitoring and sensing. Among the latest developments in this field, 3D printing of wearable devices provides an opportunity to achieve more precision for complex sensors and energy storage devices [2–4].

Additive manufacturing (AM) is a relatively new process compared to other established manufacturing processes. It is also called rapid prototyping or 3D printing (a recent and popular name) [5]. Generally, an AM process is defined as a layered manufacturing process in which a three-dimensional complex part can be fabricated by adding raw materials in a layer-by-layer manner. The process is carried out in a few steps, starting from designing the required computer-aided design (CAD) based geometry using available CAD tools, followed by its conversion to a standard tessellation language (STL) file, and slicing it according to layer thickness to ensure good part density. Then the sliced STL file is processed to design the required support structures to eliminate the chances of damage to parts due to over-hanging structures. Depending upon the layering of the raw materials and the way the materials adhere to each layer, the process is classified into different sub-categories [6, 7]. Among the additive manufacturing processes, fused deposition modeling (FDM) and direct ink writing (DIW) are the primary extrusion-based techniques for 3D printing the desired materials, as shown in figure 7.1. A preheated nozzle is used to melt the polymeric filament in the case of FDM, whereas DIW utilizes a pneumatic nozzle mechanism to extrude the ink [2, 8, 9].

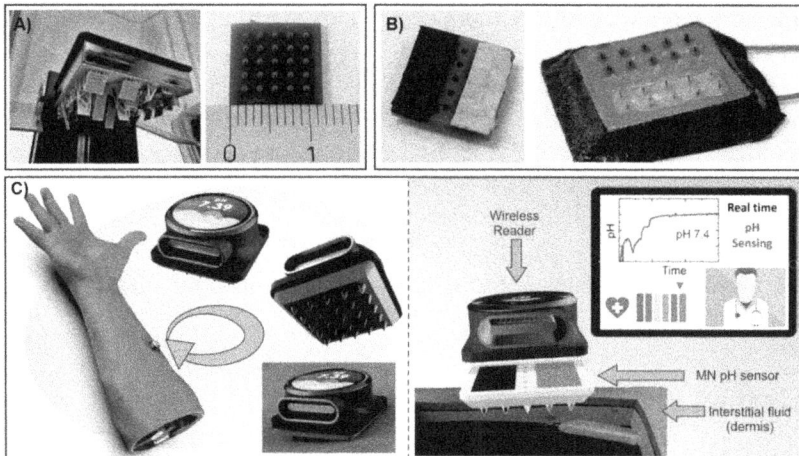

Figure 7.1. 3D-printed wearable potentiometric pH sensor. (A) Fabrication of microneedles, (B) surface tailoring of fabricated hollow microneedle arrays (front and back), and (C) conceptual illustration of pH monitoring using the 3D-printed fabricated device. (Reproduced with permission from [18]. Copyright 2023 Elsevier.)

Materials such as metals, polymer, ceramics, etc, can be easily fabricated using different processes by means of different curing, sintering, or diffusion methods. Each process has its own unique approach to depositing raw materials and finally give the desired shape and properties to the fabricated parts [10]. Several parameters, such as pressure, strain, force, and displacement, play a crucial role in developing critical wearable applications. Researchers have demonstrated biosensors, pressure sensors, strain sensors, etc, on wearable substrates for monitoring the parameters [11]. Lithography, cut-and-paste, lamination, coating, and additive manufacturing have been used to fabricate wearable devices over the years. Among these methods, additive manufacturing has proved its suitability with an ease of manufacturing for sensor development [2, 12]. Traditional manufacturing methods show several drawbacks, such as high cost of production, higher time consumption, limited flexibility of the process, etc. Additive manufacturing produces the required design with high accuracy at a lower cost than traditional methods. This method provides a higher scope for customization as per user requirements. The lower level of human interaction enables 3D printing to produce more defect-free products, with flexibility in their applications [5, 13]. In this chapter, we critically examined the applications of 3D printing in wearable devices for the fabrication of biosensors to monitor different biomarkers using various bodily fluids and energy storage devices. The fabrication process is explained in detail along with the parameters affecting it.

7.2 3D-printed biosensing devices

The demand and need for wearable 3D-printed bioelectronics are rising with population growth. They are revolutionizing the healthcare industry by creating point-of-care models for disease monitoring. The conventional monitoring systems are time-consuming and often require developing symptoms to detect a particular disease. Hence, it is desirable to produce a preemptive model using wearable devices to monitor critical parameters on-site, in real-time, with great accuracy so that early detection may be possible [14, 15]. Moreover, personalized medicine could be achieved according to personal characteristics and health parameters. Monitoring vital parameters is essential in screening for diseases and infections. In recent times, researchers have emphasized the development of various wearable biosensors to measure health parameters, which raised the potential for advanced precision medicine. Different bodily fluids, such as sweat, urine, tears, etc, have numerous biomarkers that could be useful in detecting various diseases. Also, an imbalance of electrolytes in the human body may result in multiple heart-related and kidney diseases [16, 17]. Interestingly, researchers have developed microneedle-based technologies for continuously monitoring biomarkers using bodily fluids. Interstitial fluids are often used for detection as this provides several benefits, mainly avoidance of painful blood extraction and a less complexity matrix than blood composition, etc [18]. Wearable electrochemical biosensors are widely used to detect critical components for the rapid and early detection of diseases. 3D printing of electrochemical sensors provides an economical solution with high repeatability in the analytical readouts. Chronoamperometry, cyclic voltammetry, and square-wave voltammetry are used for electrochemical detection in these miniaturized devices [19].

Sweat is a noninvasively available bodily fluid differentiated into eccrine and apocrine sweat. Eccrine and apocrine sweat glands are widely distributed in the human body. The chemical properties, location, and response of both types of sweat are different. We can measure parameters such as sodium, potassium, and chloride concentration in the sweat, which have physiological importance [20, 21]. Also, sweat provides access to abundant biomarkers such as metabolites, biomolecules, ions, glucose, cortisol, neuropeptides, and drugs. Eccrine sweat contains electrolytes and water, which are directly excreted to the skin surface [22]. Unusual changes (imbalances) in the existing sweat components can indicate health conditions. For example, the alcohol concentration in sweat is highly correlated with blood alcohol concentration. An increase in urea concentration may indicate kidney-related diseases. Furthermore, a higher chloride concentration in sweat indicates the possibility of cystic fibrosis (CF), where sweat is mainly used for the diagnosis [23].

A 3D-printed mechanically all-inclusive integrated wearable (AIIW) patch was developed by Kim *et al*, which is used for continuous monitoring of multiple electrolytes in sweat. The fabricated device was inexpensive, customizable, and could be helpful in a multiplexed approach. The 3D printing process was optimized to incorporate 3D-printed sensors and wearable-microfluidic sample handling (WMFSH) units. Bending cycles were introduced to check the fabricated patch's electrical and mechanical stability. The *in situ* monitoring of Na^+, K^+, and Ca^{2+} ions was achieved in the sweat samples [24]. Moreover, a wearable 3D-printed glucose-monitoring electrochemical ring (e-ring) was fabricated by Katseli *et al* using three plastic electrodes of conductive material incorporated with a flexible plastic holder of non-conductive material. Gold film was electrodeposited to modify the e-ring and coupled with a miniaturized potentiostat, allowing nonenzymatic amperometric glucose analysis in sweat samples for a range of 12.5–400 $\mu mol\ l^{-1}$. The proposed design mitigates the challenges of existing technologies and offers in-house possibilities for wearable sensor developments. The fabricated ring showed high sensitivity, selectivity, and stability [25].

Temperature asymmetry is a critical parameter in diabetic foot ulcers. A wearable temperature monitoring device was developed by Beach *et al* to measure the temperature while standing and sitting. Dynamic temperature parameters were estimated from four sites on each foot to calculate the temperature variation. The temperature increase time measured by the fabricated device could be a unique biomarker related to vascularization and soft tissue biomechanics for diabetic patients. The temperature increase time is faster in the case of diabetic patients compared to healthy persons [26]. Kim *et al* developed a wearable pressure sensor for the continuous and precise measurement of physical signals for the early diagnosis of diseases. 3D-printed rigid microbump-integrated liquid metal-based soft pressure sensors (3D-BLiPS) were fabricated with enhanced sensitivity of the pressure sensors (0.158 kPa^{-1} @ 50 kPa) by concentrating the microchannels' deformations. The fabricated device was highly robust to changes in temperature and multi-direction bending. The wristband device was used to monitor the pulse signal, and a wireless heel pressure monitoring system was used to measure the heel pressure for various floor types with different body positions [27]. The pH of any bodily fluid plays an essential role in understanding the influence of biological processes. Advanced systems use microneedle-based scaffolds as patches or bandages with polyaniline coatings for pH monitoring [28]. The continuous monitoring of biomarkers provides insights into various vital parameters. Interstitial fluids could be used by microneedle-based sensors on the dermal layer to detect the multiple analytes present. 3D printing permits rapid prototyping of such devices with micronsize resolutions. 3D-printed microneedle patches were fabricated for a pH monitoring application by Parrilla *et al.* Conductive ink was used to modify hollow needles to demonstrate a potentiometric sensor. The incorporation of the electrode inside the hollow microneedle was a crucial step to avoid electrode damage during operation, as shown in figure 7.1(B) [18]. Moreover, triboelectric nanogenerator (TENG) based self-powered sensors have captured attention due to their superior advantages in machine–human interfaces. A toroidal triboelectric sensor with fabric electrodes was fabricated on a MXene/Ecoflex composite material. The MXene composite enhanced triboelectric performance, surface charge density, and flexibility. 3D printing of a pyramidal structure reduced the fabrication complexity and improved the TENG operationality [29].

Furthermore, double-network (DN) hydrogels have been explored widely for wearable applications due to their versatility and flexible crosslinked structure. DN hydrogel reinforced with graphene oxide (GO) and crosslinked by polyacrylic acid (PAA) was prepared by Wang *et al* to enhance mechanical performance and self-healing efficiency. 3D printing of shear-thinning ink was performed by adjusting the degree of polymerization. The prepared gel network showed exceptional stretchability to 2500% with higher stiffness (Young's modulus: –753.667 kPa). The self-healed hydrogel maintained –88.02%, 86.93%, and 87.47% of its tensile strain, toughness, and tensile strength due to reversible hydrogen bonding and ionic interactions. The developed DN hydrogel was employed for wearable electronic biosensors and ammonia detection [30].

7.3 3D-printed energy storage devices

The wearable electronics and sensor market are proliferating with the increased commercialization of flexible electrochemical energy storage (EES) devices. These include electrochemical supercapacitors, batteries, etc, with high energy and power density. Such storage methods could be helpful for electronic skin (e-skin) applications [30]. These systems mainly consist of current collectors, electrodes, electrolytes, etc. Traditional methods, such as lithographic and screen printing, have been employed to fabricate these electrodes. However, 3D printing is considered a low-cost, rapid, controllable, and customizable alternative approach for fabricating electrodes [31]. The basic steps for fabricating EES devices include ink preparation, deposition, encapsulation packing, and electrolyte filling. These steps are time- and energy-consuming in nature. However, 3D printing of such devices could be a more convenient and user-friendly approach. 3D printing allows direct writing of the conductive ink with the required design and controlled chemical composition per the user's requirements. The functional ink must retain its shape over multiple layers while extruding continuously on the substrate. Printing speed and viscosity could be the critical parameters during 3D printing. In addition to these parameters, the solvents, fillers, and binder material also play a crucial role in the conductivity and capacitance of the devices. The filler material could be metallic, organic, or ceramic, depending on the specific application [32, 33].

Moreover, researchers have developed a conductive ink using nanomaterials and their composites. The large surface area on the nanomaterials offers numerous sites for ionic reaction and intercalation. Nanomaterials or their composites are premixed before the actual 3D printing. Their addition enhances the electrical conductivity, surface functionalities, and mechanical properties. Polymeric binder materials facilitate the filler material's homogeneous dispersion into the ink to tune the viscosity. It holds the ink together after the evaporation of the solvent. The choice of binder is dependent on the properties of the fillers [34]. The molecular weight, conductivity, viscosity, and surface tension also play a crucial role in binder selection, as shown in figure 7.2. Another vital component in ink is the solvent.

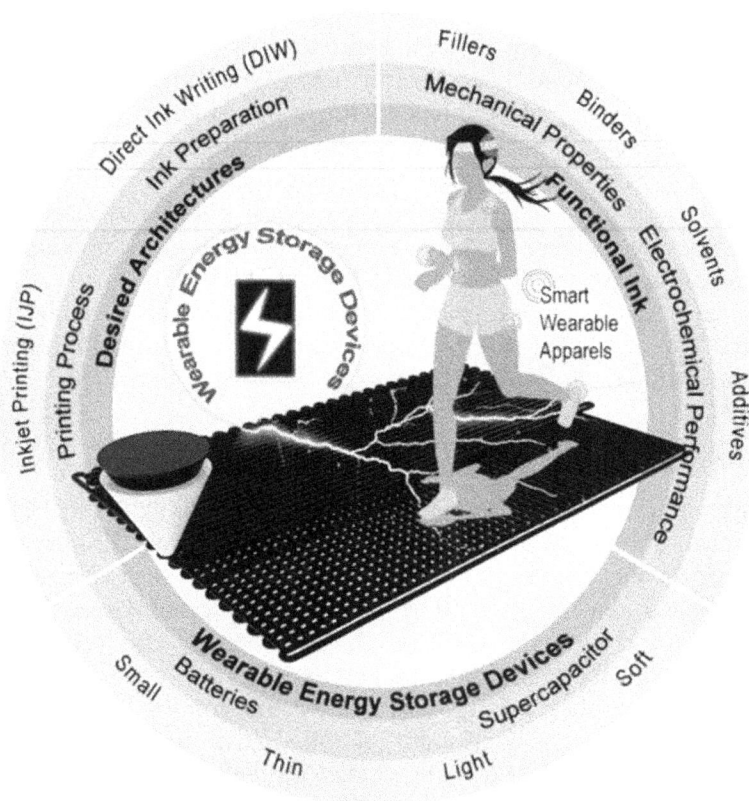

Figure 7.2. Components of 3D-printed wearable energy storage devices. (Adapted with permission from Wiley from [40]. Copyright 2022 the authors.)

The solvent used in the ink preparation should be suitable for the polymeric binders and nanofiller material. The ideal solvent for the ink can be determined using the Hillenbrand and Hansen solubility parameters, which includes the cohesive energies between the solvent and other ink components [35].

Rechargeable lithium-ion batteries and supercapacitors are considered robust players in the portable electronics sector. Conventional batteries are mechanically robust but lack flexibility and breathability, hence they cannot meet the demand created by the flexible electronics and wearable sensor industries. Therefore, considerable efforts have been undertaken to produce flexible batteries and super-capacitors to mitigate these issues using 1D fibers, 2D papers, and 3D frameworks [36]. Transition metal oxides, conductive polymers such as polypyrrole (PPy), poly (3,4-(ethylenedioxy) thiophene) (PEDOT), and polyaniline (PANI) showed superior results for pseudocapacitance materials. This faradic ion intercalation process is highly reversible with the charging–discharging process. Psuedocapacitive super-capacitors provide a higher energy density with a lower cyclic life than electrical double-layer capacitors [37, 38]. 3D-printed flexible lithium-ion batteries attracted significant attention, particularly in electrode preparation strategies. Wang et al demonstrated 3D-printed lithium-ion batteries using $LiFePO_4$ (LFP) and Li_2TiO_3 (LTO) materials with a poly(vinylidene fluoride) (PVDF) binder and a carbon nanotube (CNT) conductive additive. Both fabricated electrodes exhibited good cyclic stability and capacity. Fiber-based batteries can find applications in smart textiles and woven fabrics [39].

Carbon-yarn-based supercapacitors were developed by Carvalho et al using cellulose-based ionic hydrogels as an electrolyte source. The fabricated super-capacitor retained 92% capacitance after 10 000 consecutive charge–discharge performances. One-dimensional fiber-shaped supercapacitors were fabricated using commercial carbon fiber yarn. Twisted cotton yarn was utilized as a separator in the fabrication. Later, carbon fibers were aligned in a 3D-printed ABS mold where the threads were stretched and held together. The cellulose-based ionic hydrogel was impregnated, where carbon fiber yarn was immersed in the electrolyte solution. An in $situ$ regeneration process was achieved using glacial acetic acid. Increased energy density and specific capacitance were observed with a decrease in equivalent series resistance at higher relative humidity values [41]. Furthermore, a multilevel bonded configuration was developed by Yang et al using a 3D printing approach. In $situ$ polymerization of a polyacrylamide (PAAm)-based electrolyte was achieved where alginate chains were coupled to the calcium ions. Such coupling introduced chemical bonding and enhanced the material's tensile properties. Later, an MXene-based supercapacitor was fabricated with a maximum areal capacitance of 2.7 F cm^{-2} and reversible stretchability. Arc-shaped negative Poisson's ratio (NPR) patterns were created using 3D printing technology. A homogeneous mixture of sodium alginate and MXene was prepared and used as a viscous ink. During the process, sodium alginate acted as a hydrogen bonding agent with the hydrophilic $Ti_3C_2T_x$ nano-sheets, facilitating the crosslinked network using shear-thinning ink. An optimized ink preparation eliminated organic solvents, making it an eco-friendly process [42]. Wearable carbon fiber-based asymmetric supercapacitors were developed by Ma *et al* using aqueous MXene/polyaniline ink. MXene sediments were used directly as the ink material. The fabricated supercapacitor was compatible with a wide voltage window and utilized to power a pressure sensor for signal and motion monitoring. PVA/H$_2$SO$_4$ gel was used as an electrolyte along with MXene/polyaniline ink-based electrodes and a nonwoven separator. The synthesized material showed a prolonged cycle life and high capacitance [43]. A novel wearable supercapacitor was fabricated by Chen *et al* using 3D graphene-polypyrrole on iron (Fe)-doped MnCo$_2$O$_4$ nanoarray (3D G-PPy@Fe-MnCo$_2$O$_4$) electrodes. The iron doping mechanisms were investigated using molecular level density functional theory (DFT). The total density of the state became continuous when the partial substitution of iron into the MnCo$_2$O$_4$ lattice occurred at the Fermi energy level. Due to the above effects, the electron transfer rate was enhanced greatly. Further, a lightweight and flexible 3D G-PPy network was prepared to grow Fe-doped MnCo$_2$O$_4$ nanowire arrays vertically. This interconnected structure provided a large surface area with increased mechanical properties [44].

In a recent study, Li *et al* developed a 3D-printed micro-supercapacitor using PEDOT:PSS/MXene composite gels. MXene was mixed in ethylene glycol and an aqueous mixture of PEDOT:PSS to prepare conductive ink. 3D printing of these materials offered the appearance of hierarchically porous quinoid PEDOT structures with an optimal electron transport mechanism. The fabricated micro-supercapacitors exhibited high cyclic stability of 6000 cycles with 83% capacitance retention [45]. A symmetrical and quasi-solid-state supercapacitor was fabricated using a 3D printing approach using aqueous MXene sedimented ink by Yuan *et al*. The MXene sediment concentration was varied to achieve the required viscoelasticity. The fabricated supercapacitor showed a high areal energy density of 207.81 μWh cm^{-2} with superior cyclic stability. The flexibility and bendability demonstrated by the device proved its usefulness for wearable applications such as electromagnetic shielding, sensors, and as a power source [46].

7.4 Future prospects and conclusion

Although considerable developments have been made in the field of 3D printing in the manufacturing sector, its application in the field of 3D printing textiles is relatively in its infancy. The novel approach offered by 3D printing fabrics has the potential to significantly improve the technical capabilities of wearable sensors, provided the challenges facing the existing technologies are successfully overcome. One of the limitations of 3D-printed fabrics is their stiffness relative to conventionally made materials that limits their wearability. Since wearable sensors remain in direct contact with the body and are often worn for a long time, they should be comfortable for the user. Efforts should be made in modifying the properties of materials used in 3D printing to reduce their stiffness and make them more comfortable for users. Nature-inspired approaches offer promise in addressing issues concerning material properties without compromising performance. The development of textiles with a chain-mail structure could be explored to provide improved structural design. 3D printing could also enable the fabrication of multidimensional hierarchical bionic structures. The design of 3D-printed electrically conductive textiles should be further investigated so that a bioelectrical harvesting fabric could be developed for making sustainable sensing devices. Bioresorbability is another aspect that should be considered while developing wearable sensors, especially if they are to be used inside the body. These materials should be suitable for incorporating wireless communications. The selection of bioresorbable material having these capabilities as well as being compatible with 3D printing is a challenging task. Triboelectric nanogenerator-based wearable devices would open avenues for simultaneous energy harvesting and its application in meeting energy requirements for running miniature sensors.

The user acceptance of innovative technologies is a crucial aspect for widespread commercialization. 3D printing can allow the fabrication of highly customized wearable sensors. The integration of progress in 3D printing and modern information technologies can enable the development of advanced wearable sensors. The incorporation of the Internet of Things (IoT) can vastly expand the capabilities of such sensors. Real-time monitoring using self-powered wearable sensors can aid in improving health systems and providing timely healthcare support. Further efforts should aim at imparting multifunctional capabilities to wearable sensors.

References

[1] Bassoli E, Gatto A, Iuliano L and Violante M G 2007 3D printing technique applied to rapid casting *Rapid Prototyp. J.* **13** 148–55

[2] Liu C *et al* 2018 3D printing technologies for flexible tactile sensors toward wearable electronics and electronic skin *Polymers* **10** 629

[3] Hierlemann A, Brand O, Hagleitner C and Baltes H 2003 Microfabrication techniques for chemical/biosensors *Proc. IEEE* **91** 839–63

[4] Xu J, Fang Y and Chen J 2021 Wearable biosensors for non-invasive sweat diagnostics *Biosensors* **11** 245

[5] Lee J Y, An J and Chua C K 2017 Fundamentals and applications of 3D printing for novel materials *Appl. Mater. Today* **7** 120–33

[6] Shahrubudin N, Lee T C and Ramlan R 2019 An overview on 3D printing technology: technological, materials, and applications *Procedia Manuf.* **35** 1286–96

[7] Praveena B A, Lokesh N, Buradi A, Santhosh N, Praveena B L and Vignesh R 2022 A comprehensive review of emerging additive manufacturing (3D printing technology): methods, materials, applications, challenges, trends and future potential *Mater. Today Proc.* **52** 1309–13

[8] Cardoso R M *et al* 2020 Additive-manufactured (3D-printed) electrochemical sensors: a critical review *Anal. Chim. Acta* **1118** 73–91

[9] Davoodi E *et al* 2020 3D-printed ultra-robust surface-doped porous silicone sensors for wearable biomonitoring *ACS Nano* **14** 1520–32

[10] Jandyal A, Chaturvedi I, Wazir I, Raina A and Ul Haq M I 2022 3D printing—a review of processes, materials and applications in industry 4.0 *Sustainable Oper. Comput.* **3** 33–42

[11] Kalkal A *et al* 2021 Recent advances in 3D printing technologies for wearable (bio)sensors *Addit. Manuf.* **46** 102088

[12] Bishop G W 2016 3D printed microfluidic devices *Microfluidics for Biologists: Fundamentals and Applications* (Cham: Springer) pp 103–13

[13] Du Y *et al* 2021 Hybrid printing of wearable piezoelectric sensors *Nano Energy* **90** 106522

[14] Syedmoradi L, Norton M L and Omidfar K 2021 Point-of-care cancer diagnostic devices: From academic research to clinical translation *Talanta* **225** 122002

[15] Darabi M A *et al* 2017 Skin-inspired multifunctional autonomic-intrinsic conductive self-healing hydrogels with pressure sensitivity, stretchability, and 3D printability *Adv. Mater.* **29** 1700533

[16] Kumar S *et al* 2021 Aspects of point-of-care diagnostics for personalized health wellness *Int. J. Nanomed.* **16** 383–402

[17] Cardoso R M *et al* 2020 3D-Printed graphene/polylactic acid electrode for bioanalysis: biosensing of glucose and simultaneous determination of uric acid and nitrite in biological fluids *Sens. Actuators* B **307** 127621

[18] Parrilla M, Vanhooydonck A, Johns M, Watts R and De Wael K 2023 3D-printed microneedle-based potentiometric sensor for pH monitoring in skin interstitial fluid *Sensors Actuators* B **378** 133159

[19] Algov I, Feiertag A, Shikler R and Alfonta L 2022 Sensitive enzymatic determination of neurotransmitters in artificial sweat *Biosens. Bioelectron.* **210** 114264

[20] Singh A *et al* 2021 Recent advances in electrochemical biosensors: applications, challenges, and future scope *Biosensors* **11** 336

[21] Yu H and Sun J 2020 Sweat detection theory and fluid driven methods: a review *Nanotechnol. Precis. Eng.* **3** 126–40

[22] Ouyang M *et al* 2021 A review of biosensor technologies for blood biomarkers toward monitoring cardiovascular diseases at the point-of-care *Biosens. Bioelectron.* **171** 112621

[23] Hauke A *et al* 2020 Screen-printed sensor for low-cost chloride analysis in sweat for rapid diagnosis and monitoring of cystic fibrosis *Biosensors* **10** 123

[24] Kim T, Yi Q, Hoang E and Esfandyarpour R 2021 A 3D printed wearable bioelectronic patch for multi-sensing and *in situ* sweat electrolyte monitoring *Adv. Mater. Technol.* **6** 2001021

[25] Katseli V, Economou A and Kokkinos C 2021 Smartphone-addressable 3D-printed electro-chemical ring for nonenzymatic self-monitoring of glucose in human sweat *Anal. Chem.* **93** 3331–6

[26] Beach C, Cooper G, Weightman A, Hodson-Tole E F, Reeves N D and Casson A J 2021 Monitoring of dynamic plantar foot temperatures in diabetes with personalised 3D-printed wearables *Sensors* **21** 1717

[27] Kim K *et al* 2019 Highly sensitive and wearable liquid metal-based pressure sensor for health monitoring applications: integration of a 3D-printed microbump array with the micro-channel *Adv. Healthcare Mater.* **8** 1900978

[28] Proksch E 2018 pH in nature, humans and skin *J. Dermatol.* **45** 1044–52

[29] Zhang S *et al* 2023 3D printed smart glove with pyramidal MXene/Ecoflex composite-based toroidal triboelectric nanogenerators for wearable human-machine interaction applications *Nano Energy* **106** 108110

[30] Wang Y *et al* 2019 Tough but self-healing and 3D printable hydrogels for E-skin, E-noses and laser controlled actuators *J. Mater. Chem.* A **7** 24814–29

[31] Rengier F *et al* 2010 3D printing based on imaging data: review of medical applications *Int. J. Comput. Assist. Radiol. Surg.* **5** 335–41

[32] Camargo J R, Silva T A, Rivas G A and Janegitz B C 2022 Novel eco-friendly water-based conductive ink for the preparation of disposable screen-printed electrodes for sensing and biosensing applications *Electrochim. Acta* **409** 139968

[33] Skylar-Scott M A, Gunasekaran S and Lewis J A 2016 Laser-assisted direct ink writing of planar and 3D metal architectures *PNAS* **113** 6137–42

[34] Liu L, Feng Y and Wu W 2019 Recent progress in printed flexible solid-state supercapacitors for portable and wearable energy storage *J. Power Sources* **410** 69–77

[35] Kim J, Kumar R, Bandodkar A J and Wang J 2017 Advanced materials for printed wearable electrochemical devices: a review *Adv. Electron. Mater.* **3** 1600260

[36] Deng Z *et al* 2017 3D ordered macroporous MoS2@C nanostructure for flexible Li-ion batteries *Adv. Mater.* **29** 1603020

[37] Yeen L X, Wen F H and Jing K H 2020 Characterization of electrically conductive hydrogels —polyaniline/polyacrylamide and graphene/polyacrylamide *Mater. Sci. Forum* **977** 59–64

[38] Pirovano P *et al* 2020 A wearable sensor for the detection of sodium and potassium in human sweat during exercise *Talanta* **219** 121145

[39] Wang Y *et al* 2017 3D-printed all-fiber Li-ion battery toward wearable energy storage *Adv. Funct. Mater.* **27** 1703140

[40] Zhu Y *et al* 2022 A focus review on 3D printing of wearable energy storage devices *Carbon Energy* **6** 1242–61

[41] Carvalho J T, Cunha I, Coelho J, Fortunato E, Martins R and Pereira L 2022 Carbon-yarn-based supercapacitors with *in situ* regenerated cellulose hydrogel for sustainable wearable electronics *ACS Appl. Energy Mater.* **5** 11987–96

[42] Yang J *et al* 2022 3D-printed flexible supercapacitors with multi-level bonded configuration via ion cross-linking *J. Mater. Chem.* A **10** 16409–19

[43] Ma J *et al* 2022 Wearable fiber-based supercapacitors enabled by additive-free aqueous MXene inks for self-powering healthcare sensors *Adv. Fiber Mater.* **4** 1535–44

[44] Chen Z *et al* 2022 Flexible self-supporting 3D electrode based on 3D graphene-PPy@Fe-MnCo2O4 nanostructure arrays toward high-performance wearable supercapacitors *ACS Appl. Energy Mater.* **5** 5937–46

[45] Li L *et al* 2023 Direct-ink-write 3D printing of programmable micro-supercapacitors from MXene-regulating conducting polymer inks *Adv. Energy Mater.* **13** 2203683

[46] Yuan M *et al* 2023 3D printing quasi-solid-state micro-supercapacitors with ultrahigh areal energy density based on high concentration MXene sediment *Chem. Eng. J.* **451** 138686

IOP Publishing

3D Printed Smart Sensors and Energy Harvesting Devices
Concepts, fabrication and applications
Sanket Goel and Sohan Dudala

Chapter 8

3D printing technology for next-generation energy harvesters

Suresh Balpande, M Junaid Khan and Gajanan Nikhade

The energy transition is one of society's biggest concerns and thus scientists are eager to tackle this problem. To accomplish a wise transition to a sustainable future energy scenario, energy harvesting technologies, in particular piezoelectric nanogenerators (PENGs) and triboelectric nanogenerators (TENGs), must be studied and deployed. Additive manufacturing increases the prospect of highly efficient and intelligent technologies to boost the competitiveness of sustainable energy alternatives over fossil fuels. Next-generation energy harvesters could be created by incorporating 3D printing (3DP) technology to simplify the development process. 3D-printed energy harvesters can perform and operate more effectively by following the advice in this chapter.

8.1 Introduction

Nuclear energy has substantial start-up costs and safety difficulties. Hydroelectric power plants are cheap but have a short lifespan and are rarely used due to geopolitical difficulties and unpredictable rain. Renewable energy sources are growing. Electromagnetic, electrostatic, thermal, and kinetic motion are all ways to gather energy. As technology advances, energy demand will rise, conventional energy sources will become scarcer, and environmental concerns will grow [1]. Energy harvesting from renewable sources has been accelerated due to economic and environmental limits on the world's conventional energy resources [2].

Therefore, cutting-edge technology should improve green energy equipment efficiency. This has enhanced energy-collecting applications [3]. Most people carry multiple portable electronics. Batteries are the most prevalent way to power electronic gadgets, yet they have limited and unpredictable lifespans, require the end-of-life disposal of hazardous chemicals, and increase recycling. Generators could turn mechanical energy into electricity that is used to power small gadgets.

doi:10.1088/978-0-7503-5351-9ch8
8-1

Piezoelectric, triboelectric, pyroelectric, thermoelectric, capacitive with electrets, radiofrequency waves, and triboelectric nanogenerators are common generators [4, 5]. Photovoltaic energy has been used to a great extent and proved to be large-scale energy source. The use of electromagnetic generators (EMGs) [6], piezoelectric nanogenerators (PENGs), and triboelectric nanogenerators (TENGs) has generated a lot of interest in mechanical energy harvesting. PENGs can convert minute physical deformations into energy in self-powered, small-scale devices, in contrast to EMGs, which are based on Faraday's law of electromagnetic induction and are suitable for large-scale power generation.

Conventional TENGs convert energy cost-effectively, cleanly, and sustainably using triboelectric effects and electrostatic induction. TENGs are lightweight, compact, and offer a choice of materials and frequencies. A thermoelectric generator transfers thermal energy from temperature gradients into electricity, while a pyroelectric generator does the opposite [7]. An energy harvester must produce power at least on the order of milliwatts. The potential output of various energy harvesting devices is shown in table 8.1. which proves the feasibility of energy harvesting technology [8]. Machine vibrations, body motions, wind-converted motions, and ocean waves are prominent source of vibrations and motions, as shown in the figure 8.1(a).

The power consumption of electronic modules is summarized in table 8.2. Modern low-power electronics and cloud computing have increased energy harvesting while reducing the energy consumption of electronics [9]. A review found that several sensor modules operate between 0.1 and 1 W, a range that can be easily handled by energy harvesting equipment [10].

Through various transduction processes, harvesters can capture kinetic energy from their surroundings and convert it into electricity [11]. Recent advances in the low power requirements of small portable electronics have aided in the large-scale deployment of harvesters into wearable, portable, and implantable micro- and nanodevices.

Energy harvesting is also used in a somewhat different application where it is not yet as common, as shown in figure 8.1(b) [12–15]. We describe the principle of

Table 8.1. Energy harvesting techniques and projected power level.

S. No.	Level	Energy harvesting technology	Power range Low to high	Scale	Comments
1	High	Photovoltaic	1 μW–1 MW	Macro	Matured domain based on solar energy
2		Triboelectric	0.1 μW–1 MW[a]	Micro	Technology in research mode
3	Medium	Thermoelectric	10 μW–1 kW	Micro	Based on thermal energy
4		Piezoelectric	10 μW–100 W	Micro	Kinetic motion
5	Low	Pyroelectric	0.1 μW–1 mW	Micro	Based on the thermal energy gradient
6		Radiofrequency (RF) waves	0.1 μW–1 mW	Micro	Based on RF radiation energy

[a]Projected power range.

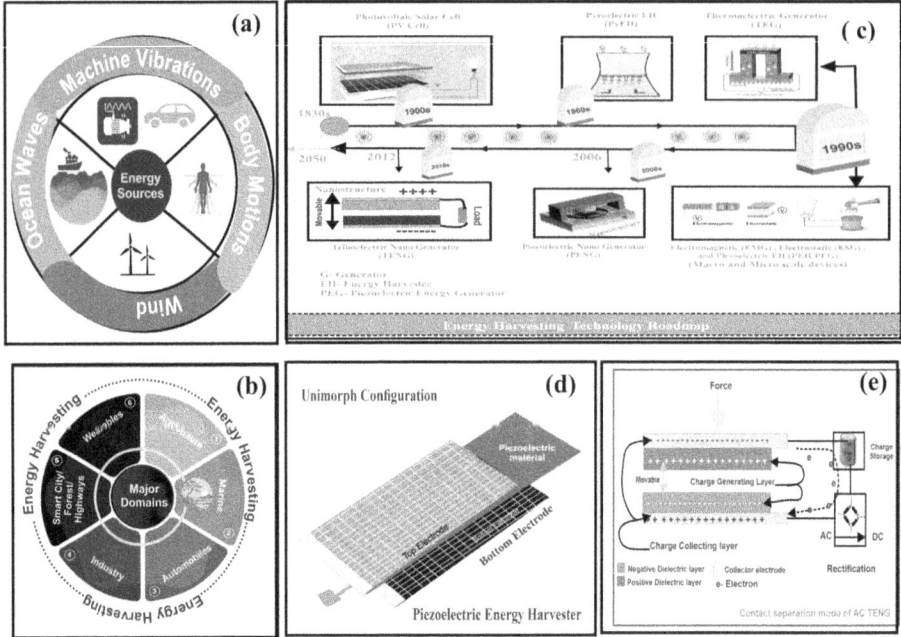

Figure 8.1. (a) Vibration energy sources. (b) Energy harvesting applications domain. (c) Roadmap for the energy harvesting technologies. (d) Typical arrangement of a PENG. (e) Typical arrangement of a TENG.

Table 8.2. Power consumption of miniaturized electronics.

S. No.	System	Power consumption	S. No.	System	Power consumption
1.	10 bit cyclic ADC	12 μW	7.	Quartz	100 nW
2.	Bluetooth transceiver	10 mW	8.	Quartz watch	5 μW
3.	Pacemaker	50 μW	9.	RFID tag	10 μW
4.	Global positioning system	100 mW	10.	Signal conditioning	50 μW
5.	Hearing aid	1 mW	11.	Watch/calculator	1 μW
6.	Hearing aid (medium)	100 μW	12.	Wireless sensor node	100 μW

obtaining energy from the human body [16] and estimate how much energy may be acquired from both movement and body heat [17, 18]. Paul-Jacques Curie and Pierre Curie discover piezoelectricity in 1880. Figure 8.1(c) shows an energy harvesting roadmap and typical applications. Scientists made the first solar energy discoveries in 1970, which enhanced environmental awareness. In 1839 Edmond Becquerel discovered the photovoltaic phenomenon, which generates an electric current when exposed to light or other radiant energy. Pyroelectric materials can turn most of the electromagnetic spectrum's energy into electrical energy. Pyroelectric harvesters were invented in the 1960s, followed by MEMS harvesters in the 1990s. PENGs used piezoelectric nanowires in the 2000s. The ability of PENGs to transform minuscule and erratic mechanical energy, i.e. vibration, wind, walking, and water waves, has drawn the most interest. TENGs can transfer mechanical energy to electrical energy and create a self-driving system. They was discovered in the year 2012 and are considered a milestone in energy harvesting [19].

8.1.1 Fundamentals of piezoelectricity

Mechanical stress increases electric charge in the piezoelectric effect, which is linearly electromechanical. A 'direct piezoelectric effect' means a material may produce an electric field in reaction to an external field. Stretched, squeezed, and pressured piezoelectric crystals deform. Charge divides some atoms, bringing them closer, and charges the electrodes. The crystals restore balance by balancing positive and negative charges at their poles. External circuits are used for balance. Basic piezoelectric harvester devices have a single interdigitated electrode or parallel plate electrodes, as shown in figure 8.1(d). Well-established piezoelectric constitutive equations link electric field and stress.

Table 8.3 lists the most prominent piezoelectric materials with critical features, including lithium niobate (LiNbO$_3$), lead magnesium niobate-lead titanate (PMN-PT),

Table 8.3. Typical piezoelectric materials [20–22].

	Critical features	PZT-5H	PMN-PT	PVDF	BaTiO$_3$	ZnO	AlN
1	Density (g cm^{-3})	7.65	8.1	1.78	6.02	5.55	3.24
2	Dielectric constant ε_r	3250	7000	6	100–1250	8.5	9
3	Young's modulus Y_{33} (GPa)	71.4	20.3	2	67	210	330
4	Mechanical quality factor Q_m	32	43–2050	3–10	1300	430–1600	2490
5	Piezoelectric charge constant d_{33} ($d_{33} = CV/F$ pC N^{-1})	100–300	2830	25	149	12.4–26.7	5
6	Piezoelectric charge constant d_{31} ($d_{31} = CV/F$ pC N^{-1})	−270	−1330	−23	−58	−5	−2
7	Electromechanical coupling factor k_{33}	0.75	0.94	0.19	0.49	0.48	0.23
8	Piezoelectric voltage constant g_{33} ($g_{33} = d_{33}\varepsilon^{-1}$ (10^{-3} Vm N^{-1}))	25	29.4	−300	11	60	50
9	Deposition temperature (°C)	⩾600	⩾600	<100	≈105	<100	250–400
10	Poling requirement	Yes	Yes	Yes	Yes	No	No

zinc oxide (ZnO), lead zirconate titanate (PZT), barium titanate (BaTiO$_3$), aluminium nitride (AlN), and polyvinylidene fluoride (PVDF) and its copolymers. PZT-based ceramics have high piezoelectric coefficients, but their stiffness, brittleness, and toxicity prohibit their use in flexible and stretchable devices. High-temperature PZT ceramics are also unsuitable. Biocompatible Pb-free ceramics such as KNbO$_3$, NaNbO$_3$, etc, can be used in implantable harvesters. Being customizable, they are the best alternative to Pb-based PZT ceramics. Polymers are well-suited for flexible devices because of their stability, biocompatibility, and flexibility.

8.1.2 Fundamentals of triboelectricity

After Wang *et al* introduced TENG in 2012, numerous researchers have improved it. The first TENG device employed Kapton and polyester. New TENG modes were introduced in 2013. TENG has two positively and negatively charged dielectric layers. The static charge and electrostatic induction cause the triboelectric effect. Charging is modeled using displacement current. Adding surface roughness increases the charge density without reducing the layer area. When one of the triboelectric layers moves, their inner surfaces come into close contact, causing a charge transfer that polarizes one side positively and the other negatively, as illustrated in figure 8.1(e). When layers are detached, opposite charges generate a voltage difference between the electrodes. Electrons travel through a load in order to reduce the voltage differences. This is repeated until the electrode potentials are equal. The voltage difference disappears when both layers are squeezed. This action links transferred charges to their layers and pulses backward, i.e. positive and negative peaks [23]. When an electrode makes contact with a dielectric, it simultaneously acts as a charge-generating layer and a charge-collecting layer. Solid and flexible electrodes, such as metals, graphene, and indium tin oxide (ITO), have been employed for many years [24–26].

Liquid–solid interface TENGs and liquid–liquid TENGs were recently introduced to scavenge water droplet energy. A fully packaged, waterproof TENG device was introduced in 2019. Their increasing power density reveals that TENGs are reliable energy harvesters and self-powered sensors that could replace batteries soon. The first triboelectric series was published in 1757 by Johan Carl Wilcke. In general, the list arranges substances from least to most likely to receive a positive charge (electron loss) and vice versa (gain electrons). Table 8.4 displays a compilation of some popular triboelectric series that have been made available by researchers. The displacement current model, which draws its inspiration from the capacitor model, is used to describe the charging process. It demonstrates that while the output power is exactly proportional to the square of the charge density, the output voltage and current of the device are directly related to the surface charge density. A frictional charge is produced as a result of the dielectric layers being brought together through a process of continuous taping [27–29]. One of the most important factors discovered is that different materials have variable densities of surface charges. Vertical contact-separation mode (VCSM), lateral-sliding mode (LSM), single-electrode mode (SEM), and free-standing mode (FSM) are the four primary modes. TENGs have several different application domains, however, the blue energy

Table 8.4. Triboelectric series with charge density.

Material	Average charge density (μC/m^2)	Type of material	Material	Average charge sensity (μC/m^2)	Type of material
Rubber	−148.20	--------	Kapton	332	++++++
Clear cellulose	−133.30		PTFE film	50 −300	
Clear polyvinyl chloride (PVC)	−117.53		Fluorinated ethylene propylene (FEP)	160−240	
Polytetrafluoroethylen e (PTFE)	−113.06		Mica	61.8	
Polycarbonate	−104.63		Float glass	40.2	
Polydimethylsiloxane (PDMS)	−102.05		Borosilicate glass	38.63	
Polyimide film (Kapton)	−92.88		PZT-5	8.82	
Polyester film (PET)	−89.44		BaTiO$_3$	1.27	
PVDF	−87.35		PZT-4	1.24	
Wood	−14.05	--	Wool	0.00014	++

harvester is the most widely used. The ocean, which covers about 70% of Earth's surface, is the planet's most powerful energy storage system. Water waves enter the triboelectric set as mechanical energy, which it subsequently captures and converts into electrical form. It is recognized as a crucial element in energy collecting [30].

8.1.3 Fundamentals of hybrid energy harvesters

The energy-sufficiency issue with a single energy harvester has lately been addressed by the use of hybrid energy harvesting technologies. In addition to obtaining energy from diverse sources, hybrid harvesting entails converting energy into electricity via several transduction techniques. Numerous researchers have demonstrated the use of a hybridized approach to energy harvesting, in which they produced energy by mixing multiple energy scavenging techniques. It uses a combination of piezo-electric–electromagnetic, piezoelectric–triboelectric, and piezoelectric–triboelectric–electromagnetic methods. These strategies could be successful shortly. This approach is similar to how piezoelectric and triboelectric harvesters can be combined to produce more effective energy scavenging and maximize their benefits while minimizing drawbacks [31].

8.1.3.1 Single-source hybrid energy harvesters

Each of the aforementioned effects has a cap on the amount of kinetic energy and electricity that can be extracted due to energy losses, which also imposes a cap on the efficiency of energy conversion. As a result, combining different kinetic energy-collecting techniques into one harvester balances out the downsides of each technique while also increasing the harvester's overall effectiveness. The most popular single-source harvesters for kinetic energy are hybrids.

8.1.3.2 Multi-source hybrid energy harvesters

In reality, a harvester may not always be able to produce enough electricity for the intended use from a single energy source that is strong and reliable. This problem is solved by structural hybridization or the use of materials with multiple functions. Ambient magnetic waves lost by electrical equipment and power transmission lines can be collected by using magnetostrictive and piezoelectric materials simultaneously [32].

8.1.4 Device development cycle

Before manufacture, one needs to utilize Multiphysics CAD to design, analyse, and optimize system performance, shape, and other critical properties. Multiphysics is a system-analysis tool. 3D printing makes three-dimensional items from digital files and will affect industrialization. A new avenue for autonomous additive manufacturing of 3D digital files, 3D printing's deposition procedures reduce waste, which is helpful when using valuable raw materials. Next-generation harvesters are made possible by the innovative additive manufacturing technology that 3D printing provides. Thermoelectric generators and fuel cells are disruptive technologies. These advances are the first in a sequence that will improve robust optimization design tools and Multiphysics simulations, and are also used to make energy devices. More applications require 3DP parts. Figure 8.2 shows the device development cycle. Once the device has been assembled, the next stage is to test it. The typical testing workbench is discussed in the electrical testing section.

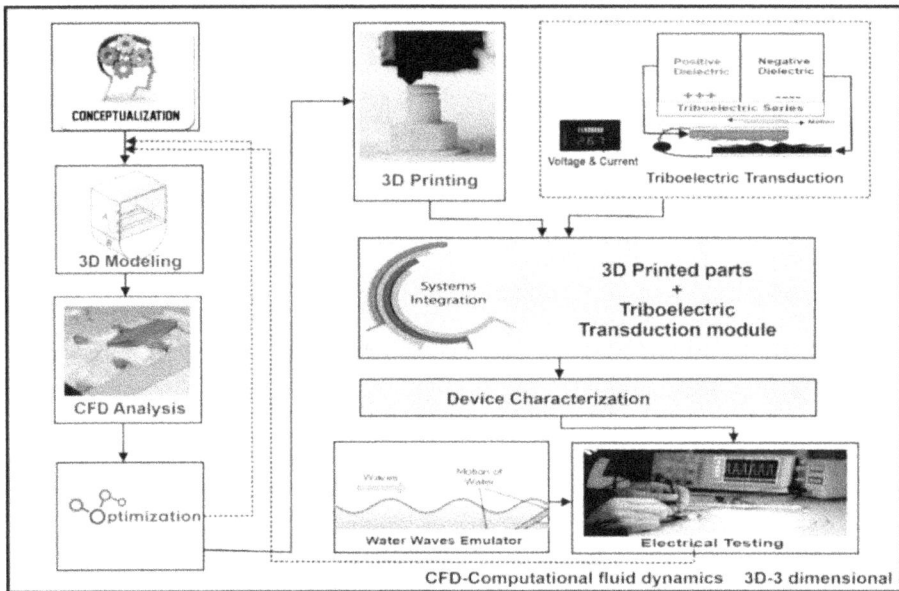

Figure 8.2. The development cycle of an energy harvester for researchers, technologists, and policymakers to use as they construct their plans for the energy industry.

Other aspects of this work are covered in the following sections. Modeling, simulation, and optimization processes are covered in the overview of the finite element analysis (FEA) tool for design and analysis in section 8.2. Section 8.3 covers the evolution of energy harvesting technology, as well as the basics, components, and 3D printing used to create the next-generation harvesters. Performance indicators and testing methodologies are covered in section 8.4. The final conclusion is presented in section 8.5. Overall, the chapters in this book will make it possible to find more realistic solutions to the problems that are now being encountered in the development of 3D-printed harvesters for the production of new harvesters.

8.2 Finite element analysis (FEA) tool for design and analysis

FEA models physical processes to reduce the need for physical prototypes, experiments, and component optimization throughout the design phase to produce better products more quickly and affordably. Mathematics is required to fully comprehend and quantify every physical process, including wave propagation, heat transfer, structural or fluid dynamics, etc. Using simulations, we can design a more sophisticated simulation that includes mechanical boundary conditions. The detailed study requires expertise in thermal, structural, and electrical design/analysis. Multiphysics simulators, i.e. COMSOL$^®$ and Ansys$^®$, are the best-known FEA tools. Although their CAD tools are not ideal (especially if SolidWorks$^®$ or Fusion 360$^®$ is used), interoperability is possible to some extent. The various stages of device modeling in the Multiphysics tool include geometry development, meshing, subdomain settings, boundary condition, selection of study, and analysis.

8.2.1 Multiphysics analysis for PENG

COMSOL$^®$ includes boundary conditions, material properties, physical interfaces, and libraries, and ANSYS$^®$ has a comparable model. This model shows a cantilever-style piezoelectric energy harvester. Materials inspired by nature such as ceramics, polymers, composites, single crystals, and biomaterials have been produced for piezoelectric generators. The stages of model development and analysis using COMSOL$^®$ are illustrated in figure 8.3(a)–(g).

8.2.2 Multiphysics analysis for TENG

The solid model is common and can be done using any FEA tool. To simulate liquid–solid interaction, you can use the ANSYS CFD simulation subset tool, which analyses fluid–fluid and fluid–solid interactions.

CFD is used to predict hydrodynamics parameters. CFD analysis is faster than conventional testing, saving design time. In our work, resistance was computed that depended on hull shape, speed, and displacement. The impact of heterogeneous hull roughness on resistance and flow characteristics has been investigated by several research groups [33]. By optimizing parent hull forms, frictional resistance is lowered. Figures 8.4(a) and (b) illustrate the various components of the hull form. The CFD analysis of the geometries, namely flat and hemispherical, has been carried out. The two-dimensional cut plot at the x–y-plane represents the velocity flow

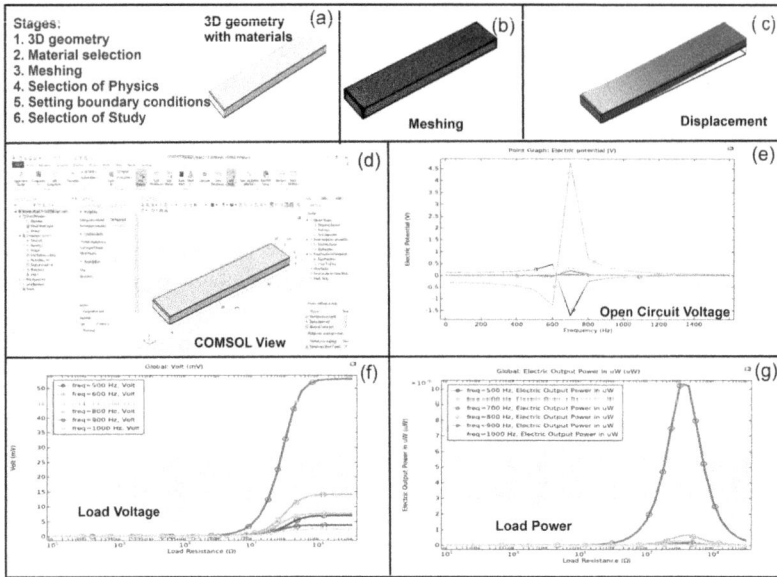

Figure 8.3. Design and analysis stages for PENG devices using the multiphysics tool.

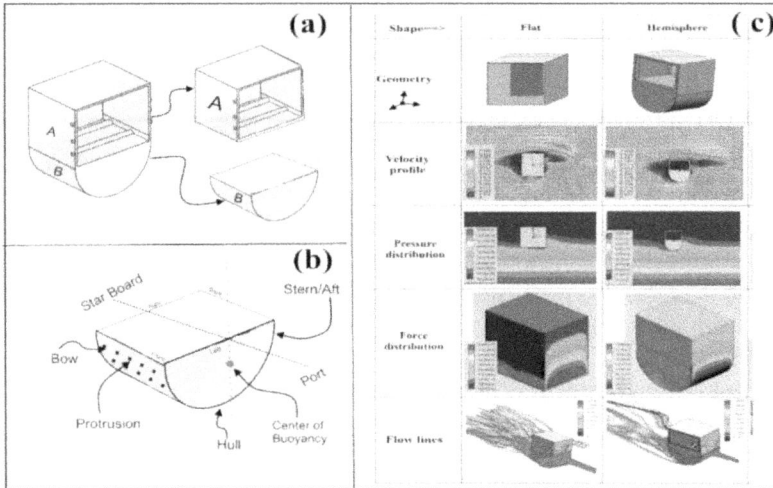

Figure 8.4. (a) The basic structure of a TENG enclosure. (b) Various parts of the 'B' subsection. (c) Modeling and CFD analysis for a TENG.

profile of all the geometries at similar flow conditions. The two-dimensional cut plot in the y-plane represents fluid (water + air) pressure distribution. The pressure force distribution on the faces of all the geometries is shown due to fluid impact at similar flow conditions. Three-dimensional flow lines represent the fluid flow trajectories over all three geometries, as shown in figure 8.4(c).

8.3 The potential of additive manufacturing (3D printing)

Cutting-edge technologies have been researched and developed to strengthen the clean energy industry. Reducing the development costs and cycles, understanding energy mixing, and the role of renewable energy in sustainability are areas of focus. Many present and future inventions will interact. Automotive devices and aircraft require light, robust materials, hence additive manufacturing has received global interest. But there are alternative methods of classification, such as organizing energy patterns, employing basic geometric algorithms, considering material composition, and analyzing support techniques [34]. Fused deposition modeling\fused filament fabrication (FDM\FFF), stereolithography (SLA), digital light processing (DLP), and selective laser sintering\melting (SLS\SLM) are the prevalent methods of consumer-grade 3D printing. Direct ink-writing (DIW) and pressure-assisted micro-syringe (PAM) are the other alternatives available for fabrication. FDM is the most popular and cost-effective form of 3D printing for energy harvesting devices.

8.3.1 3D Printed piezoelectric energy harvester

Self-powered sensors and nanogenerators based on piezoelectricity have also made great progress since 2020. The creation of piezoelectric materials may open up new possibilities for the capture of desirable green energy. KNb, PZT, $BaTiO_3$, ZnO, PVDF–trifluoro ethylene (PVDF–TrFE), lead magnesium niobate-lead titanate (PMN-PT), and others are used in the additive manufacturing of ceramics, as shown in table 8.5. Polylactic acid (PLA) is the typical filament material used in 3DP.

The 3D-printed multilayer beta phase PVDF–TrFE copolymer created by Xiaoting Yuan *et al* has a substantially higher effective piezoelectric coefficient and does not require high-temperature annealing or challenging transfer techniques. The multilayer copolymer was used to create a rugby ball shaped energy harvester with a flex tensional mechanism to demonstrate its high-power density [37]. Rui Tao

Table 8.5. 3D-printed piezoelectric harvesters [35, 36].

S. No.	3D printing technique	Piezoelectric material	Material category
1	FDM	PVDF, PVDF–TrFE, PLA mesh + poly(propene), $BaTiO_3$—acrylonitrile butadiene styrene	Thermoplastic polymers Polymer nanocomposites
2	SLA	PZT, PMN-PT, KNb, $BaTiO_3$	Photopolymers with ceramic nanoparticles
3	SLS/SLM	Polyamide/$BaTiO_3$/carbon nano tubes	Thermoplastic metal powder
4	DIW/PAM	PVDF–TrFE, $BaTiO_3$	Viscous liquid composites, sol–gel ceramics

and colleagues developed a supportive fluid-assisted multi-material extrusion-based 3D printing technique to produce flexible conformal or free-form piezoelectric composite sensors with unified electrodes. In the course of printing, two paste-like substances—piezoelectric and conductive composite inks—are extruded [38]. A 3D-printed piezoelectric nanomechanical energy harvesting device was made by Yuan *et al* for self-powered sensor applications [39]. Zhou *et al* suggested a 3D-printed flexible PENG with a non-protruding kirigami structure [40] for wearable electrical devices.

8.3.2 3D-printed triboelectric energy harvester

Casings, blades, shells, tubular components, structural frames, nano-patterns, surface changes, and friction materials have all been manufactured via 3D printing. TENG's interoperability with 3D printing technology is exceptional allowing structural supports, triboelectric materials, and electrodes, with no requirement for vacuum apparatus or high-temperature treatment, producing accurate and simple structures such as dampers, springs, and interdigitated structures, and optimizing complex design features.

3D printing makes TENGs possible and simplifies creating the enclosure's body and a provides path to next-generation triboelectric layers. In our work, SolidWorks™ was used to design the 3D structure and slices. The TENG enclosure was printed with 100% density PLA. The device includes a triboelectric set and enclosure. CFD analysis optimized the enclosure design for catching water waves. Two TENG sets coupled by a flexible material make a triboelectric set of 3 cm × 3 cm × 4 cm. Copper was used for the electrodes and PDMS and air for the dielectric layer. In this study, hemisphere-shaped (3D-printed PVA) and rectangular enclosures (acrylic) were compared. After building the TENG set and enclosure, all components were assembled, and seal with Teflon tape and Araldite.

The advancements in the 3D printing of triboelectric functional layers, mechanisms, and enclosing bodies are shown in table 8.6. Several researchers have produced TENG devices efficiently using 3D printing technology. The multi-tunnel triboelectric nanogenerator for recovering ocean wave energy is made up of polytetrafluoroethylene (PTFE) balls and two cover plates with copper electrodes. Two copper electrodes are tightly connected to the 3D-printed cover plates along the grooves, and the many semi-cylindrical grooves on the cover plates provide a tunnel for the rolling motion of the balls [48].

8.3.3 The 3D-printed hybrid energy harvester

One system may not use all the mechanical energy. To capture mechanical energy, hybridized harvesters include TENG, PENG, TEG, and EMG. These devices can complement each other's strengths. With additional input sources, such as interior light, random vibration, fluctuating thermal source, etc, a single-source harvester's output power may drop dramatically. The past ten years have seen the development of 3D-printed hybrid energy harvesters for multimode energy harvesting, including

Table 8.6. 3D-printed piezoelectric harvesters.

S. No.	3D-printed structure	Observations	References
1.	Elastic wave TENG	Sensor node application	[41]
2.	Hybrid coaxial TENG	Efficiency was 17%	[42]
3.	Bidirectional gear transmission TENG	Flywheel mass and triboelectric film	[43]
4.	Nanomechanical energy harvesting (NMEH)	0.31 W power harvested	[44]
5.	NMEH with grating disc type	Isc-18.9 A, V_{open} −231 V, and 2.2 mW power	[45]
6.	Ship hull-shaped TENG	The peak output power: 9 mW	[46]
7.	3D-printed electrodes, housing, and triboelectric parts	Sensor node application	[45]
8.	Flexible TENG	Electro-Fenton degradation system for wastewater treatment, 6 Wm^{-2}, 2 mA, and 610 V_{open}	[47]

PENG and photovoltaics, TENG and EMG, PENG and TENG, etc, as listed in table 8.7.

8.4 Test bench and device performance analysis

The electrical output performance of the constructed TENG needs to be examined using shaker-produced rotating\linear motion. An accelerometer for measuring vibration strength, a digital storage oscilloscope, a high resistance–low current electrometer, and a signal generator connected to a shaker make up the energy harvesting test bench. The user can adjust the frequency and acceleration amplitude because the entire bench is under the control of a personal computer. The most common application for TENG is based on dynamic inputs with spike-like signals. However, the traditional measuring instruments have resistance in the range of a few KΩ to 10 MΩ, and the voltage profiles were produced using a high-impedance electrometer (>1 G), which is not practical. A TENG's performance must be evaluated by measuring the voltage, including the open-circuit voltage, charges, and load output voltage. The peak and average values of the parameters are equally important. The measurement set-up and results are shown in figure 8.5.

8.4.1 Impacting parameters

TENGs and PENGs are adaptable technology used in wearables, small robotics, wireless sensors, and other applications for energy harvesting. TENGs and PENGs both transform mechanical energy into electrical power. Dimensions, material qualities, and structural geometry must all be considered carefully during the device

Table 8.7. 3D-printed hybrid energy harvesters.

S. No.	Harvesters	3D-printed structure	Observations	References
1.	EMG, TENG, and PENG	Cantilever, FDM with PLA material	Inertial sensing	[49]
2.	EMG, TENG	A magnet linked to a spring in the 3D-printed tube	Peak power outputs: EMG: 717 mW @ 600 Ω TENG: 18.9 mW @ 2 MΩ and 1.7 mW @ 6 MΩ	[50]
3.	EMG, TENG	Springless, 3D-printed frame, NdFeB N52 magnet, bobbin with the coil inside a hollow	Power: 5.41 mW @ 1.1 KΩ	[51]
4.	EMG and TENG	Wrist-wearable hybrid harvester 3D printed in ABS, hollow curved tube	EMG: Power density −5.14 mW cm^{-3} @ 49.2 Ω load TENG: 0.22 μW cm^{-3} with 13.9 MΩ	[52]
5.	EMG, TENG, and PENG	Spherical hybrid harvester	Inertial sensor for motion recognition	[53]
6.	EMG-TENG	Pendulum, printed cylindrical shell (white resin)	EMG: 523 mW @ 280 Ω TENG: 470 μW @ 0.5 MΩ	[54]
7.	EMG-PENG	Piezo-electromagnetic hybrid harvester, 3D-printed mass block with PLA	PENG: 1.28 mW peak power EMG: 30 μW and 1.31 mW for a combination	[55]

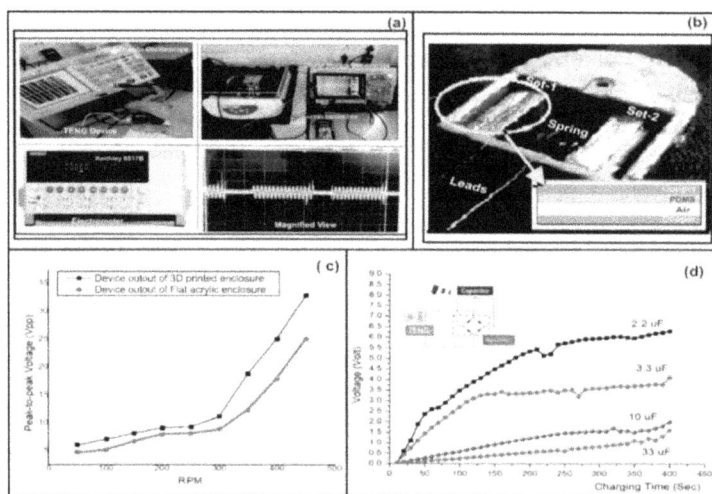

Figure 8.5. (a) Measurement set-up. (b) Assembled view of TENG. (c) Open-circuit voltage for 3D-printed and flat acrylic enclosures versus shaker RPM. (d) Capacitor profile versus charging time.

development process. During device testing, load impedance (Z_L), and excitation magnitude and frequency are managed.

(a) **Material properties**

The piezoelectric voltage coefficient, g, is regarded as a significant parameter in PENG for maximizing voltages. On the other hand, d, $k^2 \times Q_m$, $g(\tan\delta)^{-1}$, and $d \times g$ are close to the behavior of the optimum power, so these material parameters are suitable to assess the harvester's performance, where d is the piezoelectric charge constant, k^2 is the electromechanical coupling coefficient, Q_m is the mechanical quality factor, $\tan\delta$ is the loss tangent, and g_{31} is the piezoelectric voltage coefficient.

The structure of triboelectric surfaces, the contact area affected by the applied force, and the relative location of triboelectric pairs in the triboelectric series are all factors that affect triboelectric charge density (σ_T).

(b) **Structural geometry and dimensions**

This is supported by tests using triangular and rectangular beams in PENG. It turns out that in terms of curvature homogeneity irrespective of the proof mass, triangular beams perform better than rectangular ones. The other forms of beams are Pi, T, E, and rectangular, according to several scholars. The component TENG layers' thickness directly affects the polarization and propagation of their electric fields, which in turn affects the power output. Increasunf L expands the TENG layers' surface area, increasing the max output power.

(c) **Magnitude and frequency of excitation**

The magnitude and loading frequency are related to the harvested power. The piezoelectric device's capacity for producing charge drops as frequency rises. With an increase in loading frequency, the piezoelectric unit's electric energy output increases. The magnitude of excitation is also called accelcration. It can be measured using one of the sensors, i.e. an ADXL362 accelerometer.

By increasing the TENG layer's movement frequency, the power output is enhanced. The rate of transmission of these charges increases proportionately as frequency rises. Frequency enhancement strategies may be recommended to transform low input frequencies into high frequencies to acquire larger output power in real applications where the ambient mechanical motion frequency's n may be constrained by physical constraints. To successfully capture electricity, cantilever-type structures have also been employed to transform low-frequency plucking motions into high-frequency oscillations.

(d) **Load impedance (Z_L)**

This holds for all kinds of harvesters and has an inverse connection to peak power. The optimal load impedance value results in the global peak power. TENG impedance plots precisely pinpoint the fluctuation in impedance and source current and, therefore, power output, assisting in the decision-making process for the best motion settings.

8.4.2 Electrical measurements

The typical performance metrics applicable to all types of harvesters are open-circuit voltage (V_{oc}), short-circuit current (I_{sc}), load voltage (V_L), load current (I_L) for resistive/capacitive load, resonant frequency (F), and power (P). The electrical performance of harvesters is tested using a vibration source to exert a direct strain on the structure. The test bench is meant to gather energy from low-frequency hits. The application of harvesters needs to be confirmed by charging capacitors and driving loads such as LED/LCD/sensor nodes, which is easy and efficient to verify reliability.

8.4.2.1 Open-circuit voltage testing
The structure is periodically energized for various oscillating motions that somewhat resemble ocean waves. For a 3D-printed enclosure with a hemispheric bottom and an acrylic rectangular enclosure with a flat bottom, it was found that the open-circuit voltage was 32 and 24 V_{max}, respectively. As depicted in figure 8.5, all trials were carried out with an oscillating drive at speeds between 50 and 450 spins per minute. This discovery has created new possibilities for additional TENG testing with 3D-printed capacitive load enclosures.

8.4.2.2 Short circuit current and load voltage testing
TENGs that are loaded with resistive loads can function in phases. A load resistance that is extremely low can be described as a short circuit. The load resistance in region 1 is quite low, ranging from 0.1 Ω to 1 KΩ. In this region, the peak output current is lower compared to the condition of a short circuit, and the maximum voltage is nearly proportional to the load resistance. The TENG is a source of current in the regional area. In region 2, maximum current decreases as load resistance increases (between 1 KΩ and 1 GΩ) in conjunction with an increase in voltage. During the transition, the TENG will be operating at its maximum power. Inside area 3, the load resistance is more than one GΩ, and the TENG operates as a voltage source with maximal voltage saturation.

8.4.2.3 Capacitor charging profile
TENG's inherent capacitance is represented with a variable capacitor. When collecting energy from vibrating mechanical energy, such as human movement, this capacitance is minimal, leading to a high TENG output impedance. It shows that charging a load capacitor to its intrinsic capacitance maximizes power transfer. By storing the TENG in a capacitor, its usefulness as a power source might be demonstrated. By compressing and releasing dielectric layers, alternating voltage or current is produced. As seen in figure 8.5(d), the rectifier converts AC to DC which can be stored in a capacitor. This study used a capacitor to replicate a small battery. Every experiment lasted 400 s at room temperature.

A 3D-printed enclosure charges a 2.2 F capacitor to 6 V, a 3.3 F capacitor to 3.7 V, a 10 F capacitor to 1.8 V, and a 33 F capacitor to 1.2 V. The 33 F capacitor collects the greatest charge over a long charging period, while the 2.2 F capacitor charges the fastest. The capacitor's electrical energy can power Internet of Things nodes [56].

8.4.2.4 *Resonant frequency (F) and power measurement*
The resonant frequency can be tested with an fat Fourier transform enabled electrometer or manually by sweeping the input excitation frequency. The resonant frequency is a frequency's peak value. Electrometers measure and calculate power.

8.5 Concluding remarks

Recent developments in energy harvesters could lead to battery-less electronics. Multiphysics tools based on FEA help design, simulate, and optimize the device. 3D printing energy harvesters built of ceramic or polymeric materials will create new options for boosting shape, hierarchy, and material complexity, helping to address some of technology's most critical concerns, particularly at the interface level. Innovative application situations, such as those in the portable market, require previously unexplored geometries driven by topology optimization. These steps will also increase energy and power. Next-generation energy harvesters must be made possible by 3D printing useful materials, such as high aspect ratio parts and multi-material printing. Researchers can improve hybrid energy harvesters by using single and multiple energy sources.

Data availability

The corresponding author has access to the information given in this work. The source data will be made available on request.

Conflict of interest

The authors state that there are no potential conflicts of interest.

References

[1] Wu C, Wang A C, Ding W, Guo H and Wang Z L 2019 Triboelectric nanogenerator: a foundation of the energy for the new era *Adv. Energy Mater.* **9** 1802906
[2] Kahar K, Dhekekar R, Bhaiyya M, Balpande S and Kale P 2022 MEMS based energy scavenger with interdigitated electrodes *Mater. Today Proc.* **72**
[3] Chen Y *et al* 2022 Hybridized triboelectric-electromagnetic nanogenerators and solar cell for energy harvesting and wireless power transmission *Nano Res.* **15** 2069–76
[4] Sharma A and Sharma P 2021 Energy harvesting technology for IoT edge applications *Smart Manufacturing—When Artificial Intelligence Meets the Internet of Things* (London: InTech)
[5] Kahar K, Bhaiyya M, Dhekekar R, Gawande G, Balpande S and Goel S 2022 MEMS-based energy scavengers: journey and future *Microsyst. Technol.* **28** 1971–93
[6] Kumar A, Balpande S S and Anjankar S C 2016 Electromagnetic energy harvester for low frequency vibrations using MEMS *Procedia Comput. Sci.* **79** 785–92
[7] Wang Y *et al* 2021 Flexible seaweed-like triboelectric nanogenerator as a wave energy harvester powering marine Internet of Things *ACS Nano* **15** 15700–9

[8] Balpande S S, Pande R S and Patrikar R M 2016 Design and low cost fabrication of green vibration energy harvester *Sensors Actuators* A **251** 134–41

[9] Zhang R *et al* 2021 The triboelectricity of the human body *Nano Energy* **86** 106041

[10] Balpande S S, Pande R S and Patrikar R M 2021 Grains level evaluation and performance enhancement for piezoelectric energy harvester *Ferroelectrics* **572** 71–93

[11] Balpande S S, Kalambe J P and Pande R S 2019 Development of strain energy harvester as an alternative power source for the wearable biomedical diagnostic system *Micro Nano Lett.* **14** 777–81

[12] Fahad Ali A U, Hussain Z, Numan M, Fatima B, Najam ul Haq M, Majeed S and Ahmad T 2022 Triboelectric nanogenerator based on PTFE plastic waste bottle and aluminum foil *Mater. Innov.* **2** 203–13

[13] Balpande S S, Lande S B, Akare U and Thakre L 2009 Modeling of cantilever based power harvester as an innovative power source for RFID tag *Second Int. Conf. on Emerging Trends in Engineering and Technology* (Piscataway, NJ: IEEE)

[14] Gosavi S K and Balpande S S 2019 A comprehensive review of micro and nano scale piezoelectric energy harvesters *Sens. Lett.* **17** 180–95

[15] Kahar K, Dhekekar R, Bhaiyya M, Srivastava S K, Rewatkar P, Balpande S and Goel S 2023 Optimization of MEMS-based energy scavengers and output prediction with machine learning and synthetic data approach *Sens. Actuators* A **358** 114429

[16] Balpande S and Yenorkar S 2019 Optimization of energy harvester for trapping maximum body motions to power wearables *Sens. Lett.* **17** 46–54

[17] Balpande S S, Kalambe J and Pande R S 2018 Vibration energy harvester driven wearable biomedical diagnostic system *13th Annual Int. Conf. on Nano/Micro Engineered and Molecular Systems (NEMS)* (Piscataway, NJ: IEEE)

[18] Dhone M D, Balpande S and Kalambe J 2019 Energy harvester: a green power source for wearable biosensors *Sens. Lett.* **17** 55–63

[19] Balpande S S and Pande R S 2015 Design and simulation of MEMS cantilever based energy harvester-power source for piping health monitoring system *National Conf. on Recent Advances in Electronics and Computer Engineering (RAECE)* (Piscataway, NJ: IEEE)

[20] Ruize X and Kim S G 2012 Figures of merits of piezoelectric materials in energy harvesters *PowerMEMS* vol 2012 (Atlanta, GA) pp 464–7

[21] Wei H *et al* 2018 An overview of lead-free piezoelectric materials and devices *J. Mater. Chem.* C **6** 12446–67

[22] Ju M *et al* 2023 Piezoelectric materials and sensors for structural health monitoring: fundamental aspects, current status, and future perspectives *Sensors* **23** 543

[23] Rathi S, Ashtankar P, Mehta V, Anjankar S, Rathee V and Balpande S 2022 TRIBO-SIM: a parametric simulation tool for triboelectric energy generators *Int. J. Ambient Energy* **43** 7077–87

[24] Kim D W, Lee J H, Kim J K and Jeong U 2020 Material aspects of triboelectric energy generation and sensors *NPG Asia Mater.* **12** 6

[25] Balpande S, Nikhade G and Chakole P 2021 A triboelectric generator for generating electricity using wave energy *Patent Specification* India 384331

[26] Chakole P, Rathee V, Kalambe J, Kulkarni P and Balpande S 2019 Design and development of triboelectric blue energy harvester *Int. J. Eng. Adv. Technol.* **8** 1278–83

[27] Jattalwar N, Balpande S S and Shrawankar J A 2020 Assessment of denim and photo paper substrate-based microstrip antennas for wearable biomedical sensing *Wireless Pers. Commun.* **115** 1993–2003

[28] Dhone M D, Gawatre P G and Balpande S S 2018 Frequency band widening technique for cantilever-based vibration energy harvesters through dynamics of fluid motion *Mater. Sci. Energy Technol.* **1** 84–90

[29] Balpande S S and Pande R S 2016 Design and fabrication of non silicon substrate based MEMS energy harvester for arbitrary surface applications *AIP Conf. Proc.* **1724** 020099

[30] Singh S and Balpande S 2022 Development of IoT-based condition monitoring system for bridges *Sound Vib.* **56** 209–20

[31] Rodrigues C, Gomes A, Ghosh A, Pereira A and Ventura J 2019 Power-generating footwear based on a triboelectric-electromagnetic-piezoelectric hybrid nanogenerator *Nano Energy* **62** 660–6

[32] Bai Y, Jantunen H and Juuti J 2018 Hybrid, multi-source, and integrated energy harvesters *Front Mater.* **5** 65

[33] Ravenna R *et al* 2022 CFD analysis of the effect of heterogeneous hull roughness on ship resistance *Ocean Eng.* **258** 111733

[34] Williams C B, Mistree F and Rosen D W 2011 A functional classification framework for the conceptual design of additive manufacturing technologies *J. Mech. Des. NY* **133** 121002

[35] Clementi G *et al* 2022 Review on innovative piezoelectric materials for mechanical energy harvesting *Energies* **15** 6227

[36] Pei H, Xie Y, Xiong Y, Lv Q and Chen Y 2021 A novel polarization-free 3D printing strategy for fabrication of poly (vinylidene fluoride) based nanocomposite piezoelectric energy harvester *Composites* B **225** 109312

[37] Yuan X, Gao X, Yang J, Shen X, Li Z, You S, Wanga Z and Dong S 2020 The large piezoelectricity and high power density of a 3D-printed multilayer copolymer in a rugby ball-structured mechanical energy harvester *Energy Environ. Sci.* **13** 152–61

[38] Tao R, Granier F and Therriault D 2022 Multi-material freeform 3D printing of flexible piezoelectric composite sensors using a supporting fluid *Addit. Manuf.* **60** 103243

[39] Yuan X, Gao X, Shen X, Yang J, Li Z and Dong S 2021 A 3D-printed, alternatively tilt-polarized PVDF-TrFE polymer with enhanced piezoelectric effect for self-powered sensor application *Nano Energy* **85** 105985

[40] Zhou X, Parida K, Halevi O, Liu Y, Xiong J, Magdassi S and Lee P S 2020 All 3D-printed stretchable piezoelectric nanogenerator with non-protruding kirigami structure *Nano Energy* **72** 104676

[41] Tol S, Degertekin F L and Erturk A 2019 3D-printed phononic crystal lens for elastic wave focusing and energy harvesting *Addit. Manuf.* **29** 100780

[42] Wang Y *et al* 2021 An underwater flag-like triboelectric nanogenerator for harvesting ocean current energy under extremely low velocity condition *Nano Energy* A **90** 106503

[43] Zhao T, Xu M, Xiao X, Ma Y, Li Z and Wang Z L 2021 Recent progress in blue energy harvesting for powering distributed sensors in ocean *Nano Energy* **8** 2211–855

[44] Han N, Zhao D, Schluter J U, Goh E S, Zhao H and Jin X 2016 Performance evaluation of 3D printed miniature electromagnetic energy harvesters driven by air flow *Appl. Energy* **178** 672–80

[45] Seol M-L, Ivaškevičiūtė R, Ciappesoni M A, Thompson F V, Moon D-I, Kim S J, Kim S J, Han J and Meyyappan M 2018 All 3D printed energy harvester for autonomous and sustainable resource utilization *Nano Energy* **52** 271–8

[46] Min G, Xu Y, Cochran P, Gadegaard N, Mulvihill D M and Dahiya R 2021 Origin of the contact force-dependent response of triboelectric nanogenerators *Nano Energy* **83** 105829

[47] Zhu Y, Chen C, Tian M, Chen Y, Yang Y and Gao S 2021 Self-powered electro-Fenton degradation system using oxygen-containing functional groups-rich biomass-derived carbon catalyst driven by 3D printed flexible triboelectric nanogenerator *Nano Energy* **83** 105720

[48] Zhang Z *et al* 2022 Multi-tunnel triboelectric nanogenerator for scavenging mechanical energy in marine floating bodies *J. Mar. Sci. Eng.* **10** 455

[49] Han D, Shinshi T and Kine M 2021 Energy scavenging from low frequency vibrations through a multi-pole thin magnet and a high-aspect-ratio array coil *Int. J. Precis. Eng. Manuf. Green Technol.* **8** 139–50

[50] Maharjan P, Bhatta T, Cho H, Hui X, Park C, Yoon S, Salauddin M, Rahman M T, Rana S S and Park J Y 2020 A fully functional universal self-chargeable power module for portable/wearable electronics and self-powered IoT applications *Adv. Energy Mater.* **10** 2002782

[51] Salauddin M, Toyabur R M, Maharjan P, Rasel M S, Kim J W, Cho H and Park J Y 2018 Miniaturized springless hybrid nanogenerator for powering portable and wearable electronic devices from human-body-induced vibration *Nano Energy* **51** 61–72

[52] Maharjan P, Cho H, Rasel M S, Salauddin M and Park J Y 2018 A fully enclosed, 3D printed, hybridized nanogenerator with flexible flux concentrator for harvesting diverse human biomechanical energy *Nano Energy* **53** 213–24

[53] Koh K H *et al* 2019 A self-powered 3D activity inertial sensor using hybrid sensing mechanisms *Nano Energy* **56** 651–61

[54] Xie W *et al* 2021 A nonresonant hybridized electromagnetic-triboelectric nanogenerator for irregular and ultralow frequency blue energy harvesting *Research* **2021** 5963293

[55] Shi G, Chen J, Peng Y, Shi M, Xia H, Wang X, ye Y and Xia Y 2020 A piezo-electromagnetic coupling multi-directional vibration energy harvester based on frequency up-conversion technique *Micromachines* **11** 80

[56] Balpande S and Pande S 2022 Comparative evaluation of fabric yarn, polymers, and seed crust dielectrics for triboelectric energy harvesters *J. Electron. Mater.* **51** 4270–80

IOP Publishing

3D Printed Smart Sensors and Energy Harvesting Devices
Concepts, fabrication and applications
Sanket Goel and Sohan Dudala

Chapter 9

Cyber–physical system enabled 3D printed devices

P Subbulakshmi, L K Pavithra, E Manikandan, K A Karthigeyan and Inbarani

Additive manufacturing technology is going to drive the next industrial revolution, Industry 5.0, where human and machine communication is completely possible. This rapid fabrication technology allows the fabrication of prototypes for early testing of real-time devices. With the advancements in cyber–physical systems, the necessary instructions to the system and the monitoring of it can be done remotely through the cloud. The manufacturing process involves the optimization of process parameters such as temperature, scanning speed, and so on. This also can be done by integrating the processing module with the system. Machine and deep learning algorithms will be applied for the optimization of these process parameters. The entire manufacturing process can be fully automated and remotely operated using the cyber–physical system and optimization algorithms.

9.1 Introduction

Industrialization is key to improving the world economy by transforming mechanical manufacturing into intelligent manufacturing [1]. This also helps in increasing productivity. Industry 1.0 (I1.0) started with the mechanical or traditional manufacturing system, and with the invention of electricity it entered into Industry 2.0 (I2.0). The use of computers came into existence for digital manufacturing which leds to Industry 3.0 (I3.0), the third industrialization. Recently, with the use of advanced technologies such as the Internet of Things (IoT), Big-data analytics, and artificial intelligence (AI), the entire manufacturing industry achieved fully automated digital manufacturing, which is referred to as Industry 4.0 (I4.0). Robotics is a disruptive technology that is interdisciplinary in nature and involves electronics, mechanics, computer science, and so on. Robots are being used for many applications including industrial robots, service robots such as cleaning, medical surgical applications, and the manufacturing process. Since it has been applied in manufacturing it can also be used for additive manufacturing applications. There is thus a necessity for a human–robot interaction revolution, which will be called Industry 5.0 (I5.0) [2–5]. Additive manufacturing (AM) is helping in the industrial revolution I4.0 for intelligent manufacturing, and is also referred to as 3D printing technology. This technique was initially adopted for rapid prototyping of the end product for demonstration purpose. Also, there was an initial argument that AM was only suitable for low-volume products. Now AM is being utilized even for the construction of houses with the help of novel materials. It also supports the recycling of the waste materials. The basic operating principle of AM is layer-by-layer (LbL) deposition of material in a sequential manner, although there are some methods that do not follow the LbL method. The basis for the AM process is the use of computer-aided design (CAD) software for the 3D model design or the use of a 3D scanning technique [6]. The steps involved in the design of AM and its modification using machine learning (ML) for performance prediction is shown in figure 9.1. The process starts with the 3D model of the product to be fabricated. Then a process called tessellation is applied where the input 3D model is converted into an STL file, i.e. stereolithography or standard tessellation language. Next, it involves slicing the STL file formed into G-code. These two steps, tessellation and slicing, are referred to as a physical and simulated process. This is done automatically by using the proper software. G-code is a format that can be understood by the 3D printer. The 3D printer takes the G-code input file and prints the required model layer-by-layer. The entire process is now automated with the help of the I4.0 revolution [7].

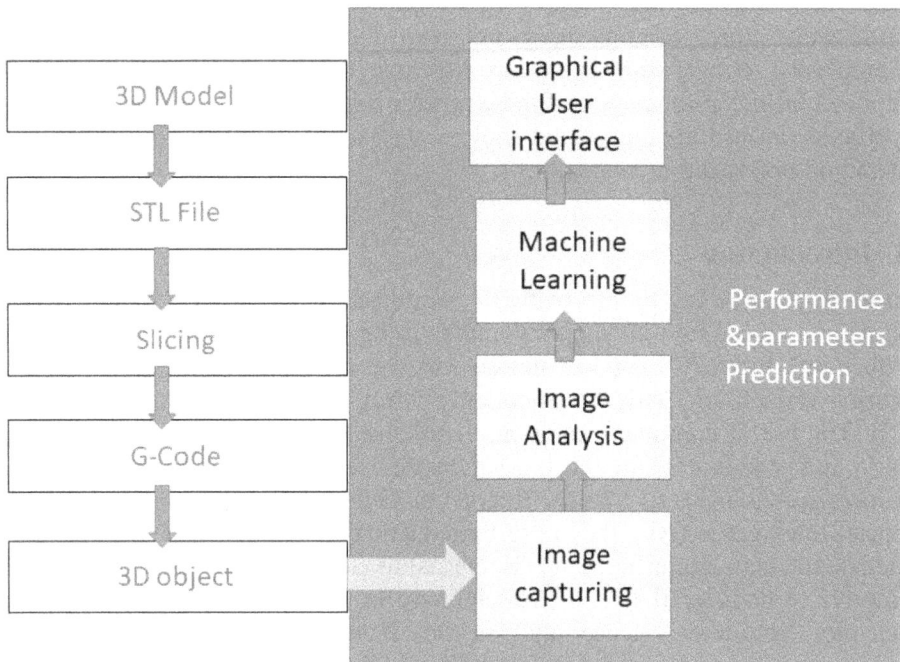

Figure 9.1. Steps involved in basic AM and the performance analysis using ML.

The major problem with the AM design flow is that, at each stage, there is a possibility of introducing vulnerabilities that could affect the entire process starting from the CAD file to accessing the 3D printer. At the early stage, the input CAD file can be altered by cyber-attack. For example, researchers edited a CAD file which was in ZIP format. The ZIP file became infected which further induced some void in the design. This kind of void may lead to catastrophic failure once it is fabricated. This is very sensitive for designs such as propeller blades, drones, etc [8]. It is also possible for STL become infected, which again induces a void in the design. With the help of the most recent industrial revolution, it is easy to access a 3D printer directly. This remote access is susceptible to cyber-attacks which alters the G-code and the final design will be altered further. It is possible to obtain G-code details by obtaining information such as vibration, power consumption, speed of the motor, magnetic field intensity, acoustics, and so on. So even without accessing the memory, the design can be altered directly.

By accessing all this information, hackers can easily achieve reverse engineering of the product. So securing the process flow model to the product is a complicated task in 3D printing. One way to enhance the security is to introduce a digital signature in the design, which has to be trackable. Hence, a CPS with proper security against these attacks will avoid the failure of AM products [9, 10].

The additive manufacturing industries should be more focused, competitive, and ready to adopt new digital technologies in their process, which is essential for continuous improvement. Many digital techniques and smart devices are being integrated into the system, which helps in collecting valuable information. This data availability enables AI and ML techniques to use cases of AM to optimize the design and its processes. The ML technique helps in optimizing the process parameters such as temperature, pressure, etc, by simply uploading the required image using the graphical user interface during the extrusion process. Many optimization algorithms are used for this optimization process for a set of 224 data points and the gradient boosting regression algorithm fits the best with better accuracy. Based on this many CPS-enabled devices have been fabricated. The 3D printed devices include a thermal sensor for indoor applications, lab-on-chip for molecular detection, pH sensing of skin, and so forth. Recently, 4D printing using shape-changing materials, mainly for cardiovascular devices, gained attention in the field of CPS-enabled AM research [11–16].

9.2 Cyber–physical systems—introduction

A brand-new category of digital systems called cyber–physical systems (CPSs) allows for unprecedented physical and computational interaction with humans. It is meant to operate more like a network of many variables with physical input and output than a stand-alone piece of technology. This sort of concept has a lot in common with the computationally intelligent sensor networks used in robotics. The ability to connect and communicate while expanding the capabilities of the physical environment through computing is a significant advancement for future technology. Cyber–physical systems are systems that include both physical (hardware) and

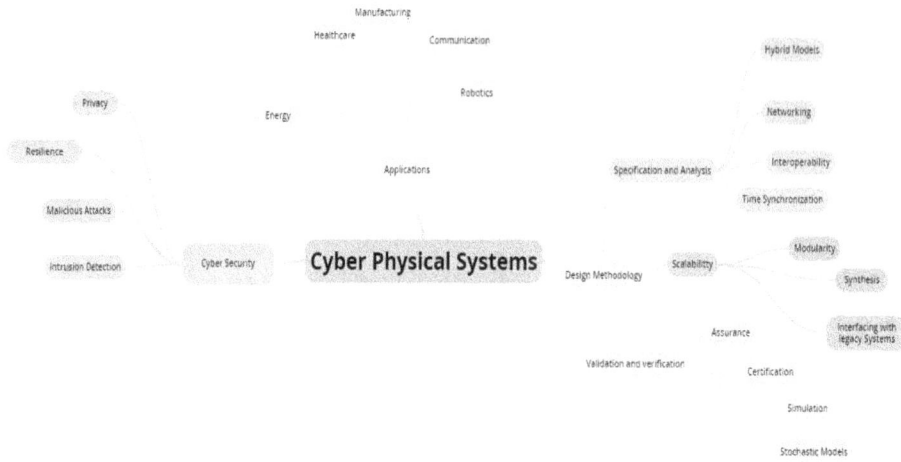

Figure 9.2. The architecture of cyber–physical systems.

software systems, as shown in figure 9.2, as well as possibly other types of systems (e.g. human systems). To produce some global behavior, these are closely connected and networked [8].

In order to communicate with the real environment as well as with sophisticated software components, CPSs frequently include hardware such as sensors, actuators, and similar embedded devices. CPSs can be found in many socially significant fields, including energy, healthcare, home automation, transportation, and agriculture.

The vast differences in design methodologies between the numerous engineering disciplines involved, such as software and mechanical engineering, provide a problem in the creation of embedded and cyber–physical systems. In today's market, when speedy innovation is seen to be vital, engineers from many disciplines must be able to work collaboratively to explore system designs, assigning responsibilities for software and physical components, and making trade-offs between them. Recent research has shown that integrating disciplines using co-simulation will enable disciplines to collaborate without introducing new tools or design processes. Robots, intelligent structures, implantable medical devices, self-driving automobiles, and aircraft that fly automatically in a regulated airspace are a few examples of cyber–physical systems.

9.3 CPS in 3D printing

Using additive manufacturing, also known as 3D printing, to create three-dimensional objects from digital models requires combining components (including liquid, powder, wires, or sheets). It is common to do this in layers. When compared to formative manufacturing (casting, forging, etc) and subtractive manufacturing, which have long histories of technological development, currently, the manufacturing sector's expansion is being constrained by a severe lack of innovation capability for the development of new products. AM, which also provides a useful method for product creation, enables the quick and efficient production of unique goods.

Additionally [17], AM could lower the technical capital and human requirements of the manufacturing sector, which supports the growth of microenterprises in the sector and mobilizes social wisdom and financial resources. AM offers excellent potential for the creation of new products and the industrial revolution, which could result in the restructuring of the manufacturing sector and encourage its transformation to intelligent manufacturing in industry.

The most important advantage of 3D printing is the release of creative flexibility in material selection and structural structure, enabling customizable shape and customized performance. The digitization of the design and production process was effectively achieved using AM technology by establishing the data flow from materials to final applications [17]. The majority of manufactured materials in various stages could be utilized as raw materials in the additive manufacturing process, which even enabled the creation and preparation of new properties and performance through the microstructure and composition of raw materials. Complex structures with a variety of functionalities can now be designed and produced in greater quantities thanks to AM, especially those with intricately curved surfaces, hierarchical lattices, and thin-wall/hollow constructions. With the aid of cutting-edge material science, design methodologies, processes, and equipment, functional and intelligent structures can also be realized on the basis of complex structures.

3D printing is a technology that produces 3D parts layers-by-layer from a polymer or a metal based material. The method is dependent on the digital data files which are being transmitted to the machine, that eventually are built as a component. 3D printing methods have become well structured, and scaled up with efficiency in recent years. The 3D printing methods provide well-defined methods and an ample amount of creative freedom for designers. They also provide several choices of push-based production systems to product managers. Since the majority of the method is dependent on the transmission of digital data files, the rise of cyber–physical systems comes into play. The 3D printing process is inherently a cyber–physical system since the digital and physical supply chains are mixed together throughout the whole product life cycle of 3D printed objects. As a result, in addition to the benefits of a CPS, the approach also adds a new class of attack vectors. We contend that current cybersecurity practises must adapt in order to handle the new class of attack vectors that threaten the AM supply chain. We also look at the characteristics of current solutions that help reduce risks and attack threats.

AM is a cyber–physical system because it combines a digital design process (the first stage of the product life cycle) with a physical production path (the CPS). With AM, the majority of design processes and their iterations may be collaborative and conducted online. Information breaches relating to supply chain locations, contracts, linked financial activities, digital usage rights, and intellectual property are possible additional cyber hazards. However, these dangers are not specific to the field of AM and might not seriously affect the caliber of the final result. The majority of cyber–physical systems are subject to these kinds of dangers. We provide a unified way of looking at both the supply chain and the AM manufacturing process chain as a CPS. With the introduction of AM technology, this amalgamation enables us to evaluate the effects of the numerous risks to the supply chain and process chain. The unified view acts as a guide for locating and putting into practise already-existing solutions and reveals security holes that need to be fixed.

Materials play a major role in all AM processes because of their innate capacity to affect performance and shape. The need for feedstock production, effective fabricator processing, effective post-processing, and the requirement for appropriate service characteristics all have an effect on the material requirements. Materials such as metals, polymers, ceramics, and natural materials are used in various AM techniques. Based on these homogeneous material systems, AM processes with heterogeneous materials, including all types of composites and multiple materials, have been successfully established in order to obtain better properties, more functions, and even customized performance, such as flame retardant polymers, direct metals, and ceramic composites. Additionally, smart materials with specific sensitive qualities, such shape memory, have been used in AM techniques to create buildings that may change in shape or performance, often known as 4D printing.

The risks associated with the physical and virtual supply networks vary. Additionally, there are different vulnerabilities and attack vectors associated with consumables such as feedstock versus fixed installations such as printers [9, 3]. Threat taxonomy in AM has been presented in various studies. The physical aspect of the AM CPS is not covered in a lot of the descriptions that are currently available because they are written from a cybersecurity standpoint. The risks and threats in the field of AM far outweigh standard cybersecurity issues such as denial of service (DoS) attacks, ransomware attacks, and hacking to sabotage, and a significant improvement in the conversation is needed to encompass other scenarios linked to the quality of the manufactured part.

The AM CPS has identified four threat categories that apply to the whole supply chain. They are:

- **Side channel attacks:** At no point has the AM supply chain been affected. The data are gathered through auxiliary systems including electricity meters, security cameras, cell phone microphones, and vibration sensors.
- **Direct sabotage:** Anywhere along the supply chain is vulnerable to this type of assault, which may even be used to insert flaws into printed parts or carry out design mutations in cloud storage.
- **Counterfeit/unauthorized production:** The capacity of the stolen digital files to generate high-quality components with an identical level of quality to the original part is a weakness unique to CPS.
- **Reverse engineering:** The attacker in this instance only interacts with the supply chain to obtain the part lawfully. Reverse engineering techniques such as computed tomography (CT) scanning, three-dimensional scanning, or building a CAD model from dimensional measurement can be used on the component.

9.4 Applications of cyber–physical systems

By incorporating sensing, computation, control, and networking, cyber–physical systems connect physical infrastructure and objects to the Internet and to one another. The National Science Foundation is a leader in sponsoring advancements in the fundamental tools and knowledge required to realize cyber–physical systems. These advancements have the power to change our society by enabling a revolution of 'smart' devices and systems, ranging from smart grids to smart cars, which combined will give rise to smart cities that can address some of the most pressing national issues [18].

9.5 Transportation and energy

In the future, we will fly in aircraft that cooperate to reduce delays and autonomous automobiles that safely interact with one another on intelligent roadways. In disaster zones, drones will check the infrastructure for damage and offer Wi-Fi. Energy for homes and businesses will be provided by a user-aware smart grid that employs sensors to monitor the environment and regulate lighting, heating, and cooling.

9.6 Healthcare

Cyber–physical systems will revolutionize healthcare delivery by making smart medical treatments and services possible. The use of prosthetic limbs and robotic surgery will help the injured and crippled recover their mobility and, eventually, even enhance human talents. Robotic surgery and bionic limbs will make personalized medical gadgets compatible, and sensors in the house will be able to detect changing health conditions.

9.7 Sustainability and environment

The use of cyber–physical systems to advance sustainability is growing. Cyber–physical systems are supporting firemen in the detection and prevention of fires in addition to improving agricultural practises and enabling scientists to reduce undersea oil leaks. In comparison to the present generation of simple embedded systems, cyber–physical system developments will result in systems that are substantially more competent, flexible, scalable, robust, safe, and usable.

9.8 Application of CPSs in 3D printing

9.8.1 Prototyping

Prototyping is the earliest and most significant use of 3D printing technology. Designers and engineers realized early on that printing their prototypes rather than having them machined would save them time and money. The prototype initially had to be delivered to a service bureau unless the business could afford one of the few highly expensive printers that were available. However, just in the last few years, printer costs have decreased substantially, and output quality has increased to the point that even inexpensive printers can generate prototype-quality components.

9.8.2 Pharmaceutical industry and biotechnology

Two unmatched value propositions are provided by AM: freedom in design complexity and customization while maintaining potential profitability. Industries have made use of the technology's promise to meet their specialized needs. The pharmaceutical manufacturing industry can generate tailored medicine doses based on a patient's mass and metabolism. In order to provide patients with anatomically correct prosthetics like bone implants, heart valves, and tracheal splints, the biotechnology industry leverages the ability to overcome design complexity limitations.

The above claims have been used in the production processes of the automobile, aerospace, and electronics sectors as well. Due to advances in material science, consumers now have access to the benefits of bespoke manufacturing, where the functionality of the product is not affected by the limitations of the fabrication process.

9.8.3 Optimization techniques in 3D printing

The design and production of an object are easier using the 3D printing technique. Automation of the design and production of the model can be achieved by the combination of a cyber–physical system and 3D printing techniques. Currently, 3D printing machines are able to produce the final product in different materials such as plastic, carbon fiber, etc. An expert-monitored 3D printing takes several trials to obtain the perfect object with plastic or any other material. Moreover, a human expert has to perform different types of assessments over the material used in the 3D printer, which is a time consuming process. Also, a huge amount of energy and resources are utilized by the printer for printing the 3D object and, hence, it will increase the material wastage and production costs of the object. In addition, an increase in material wastage and object-building time will affect some parts of the 3D printer, leading to maintenance failure.

The combination of CPSs and ML techniques is introduced in 3D printing technology to reduce the negative impact of materials caused by the environment during the object-building process. This combination increases the quality of the product printed by the 3D printing machine. Mennenga *et al* [18] enhanced CPS-enabled 3D printers by joining the product service system (PSS). The sensors in PSS-integrated CPS 3D printing captures the data to extract meaningful information and it satisfies the customer's needs accordingly. The benefit of integrating PSS with the CPS is that it gives information about the product and process that is taking place in 3D printers. Thereby it reduces the material wastage by predicting errors in the early stage of printing. Moreover, the correlation between the quantity of materials needed for printing the 3D printed parts and their properties are increased with artificial neural network (ANN) techniques. Azlin *et al* [19] used a 3D printer along with an ANN model for processing 1D, 2D, and 3D data quality checks and it reduced the amount of material involved in the 3D printing. Metals and polymer based 3D printing analysis takes place with the help of evolutionary algorithms and ANN models. ANN based 3D printing methodologies need a huge amount of data for prediction or classification. Therefore, the data collection time for the ANN based 3D printing is high. A smaller amount of data causes the over-fitting problem in classification. While designing 3D parts, the ML enabled 3D printers give a varied number of recommendations for the design, color, texture, and its size [20]. These kinds of recommendations improve the object design and give a wide range of choices for the user involved in the design task. Additionally, it gives a preview of the final object and is open to accept customized input from the user. These recommendations improve the design knowledge of the user if they lack understanding about the final output.

Like 3D object design, the object's shape also plays a vital role in 3D printing technologies [21]. The geometric shape and dimensions are essential in the final 3D printed object. Instead of Cartesian coordinates, polar coordinates are used in this process. The polar coordinate (angle) details denote the dimensions of the object. Slight variation in an angle creates enormous deviation in the final object. Therefore, Ferreira *et al* [22] used a Bayesian neural network to generate the shape deviation model for one specific 3D printing application. Then, the generated model knowledge was transferred to different 3D printing applications and parameter tuning. This will effectively identify the small variations in the shape of the object. A binary probabilistic distribution-based 3D space model along with a convolutional neural network (CNN) is proposed by Shen *et al* [23]. This model minimizes measurement variation error and helps to maintain the quality of the object. In the same way, the analysis of different kinds of materials processed by 3D printers also has equal importance in 3D printing. Decost *et al* [24] performed computer vision-based analysis of eight different types of powdered materials. The particle distribution of each powdered material is distinct. A synthetic micrograph for each material is generated by the blender. The key point for each powder material is identified and organized as a dictionary using the bag of visual words (BOVW). After this, the shift invariant texture features around the identified vital points are extracted. Then, support vector machine-based multi-classification is carried out over the feature vector, which provides 89% accuracy. The classification time and the number of feature points are directly proportional to each other, thus, Vrabel *et al* [25] took the first four dominant principle components for material analysis. One of the properties of the powder consistency is effectively classified by Valente *et al* [26] using the decision tree classification model.

The printing process of 3D printers is monitored using the visual data captured by digital cameras. The dataset containing images can be used to monitor a powder bed fusion (PBF) process. They experimented with the monitoring process using machine and deep learning techniques. In the machine learning technique, several feature vectors were extracted from the sequence of images captured in the fusion monitoring process. Principle components analysis (PCA) is applied over the images, and a few principle components were considered for the support vector machine (SVM) classification, and it achieved around 90% accuracy. On the other hand, by extracting the features from the given image, the CNN model classified the given image data and reached approximately 92% accuracy.

The typical pattern within a set of images obtained by a digital camera can be classified by a deep convolutional neural network (DCNN) [27]. The droplet phenomena of the liquid metal jet printing (LMJP) process are monitored and controlled by the neural networks. Here, the flood fill algorithm is considered for extracting the features from the image. The drive voltage of the LMJP is controlled and adjusted by properties such as satellite, ligament, volume, and speed for jet stabilization [28].

9.8.4 Application of ML and CPSs in 3D printing

A machine learning and cyber-physical system in combination in dental 3D printing is shown in figure 9.3. Recent developments in technology have brought significant changes in the healthcare industry and 3D printing is one such important technology. The use of 3D printing technology in healthcare has improved efficiency and time management. Its medical applications include dentistry, anatomical models, pharmaceuticals, medical devices, and tissues/organs. In this study, we focus on the dental field involving modern technology [29, 30].

Crowns and dental implants are the most frequently created models using 3D printing technology in dentistry. Information on the size, shape, color, position, and design of the model to be printed is collected and given as input to the 3D printer which produces the precise model accordingly. This process is efficiently achievable using machine learning and deep learning models. Use of these modern learning techniques yields a better result equivalent to a natural tooth.

Stereolithography, selective lasers, and binder jetting are some of the main 3D printing technologies used to perform the task. This technology is accompanied by CNNs and machine learning models for achieving accuracy and efficiency of the 3D model. The use of 3D printing is prone to cyber threats and it is essential to protect the process. Hackers might access it to steal vital information and resources, hence the role of the cyber–physical system, which safeguards the printer against malware

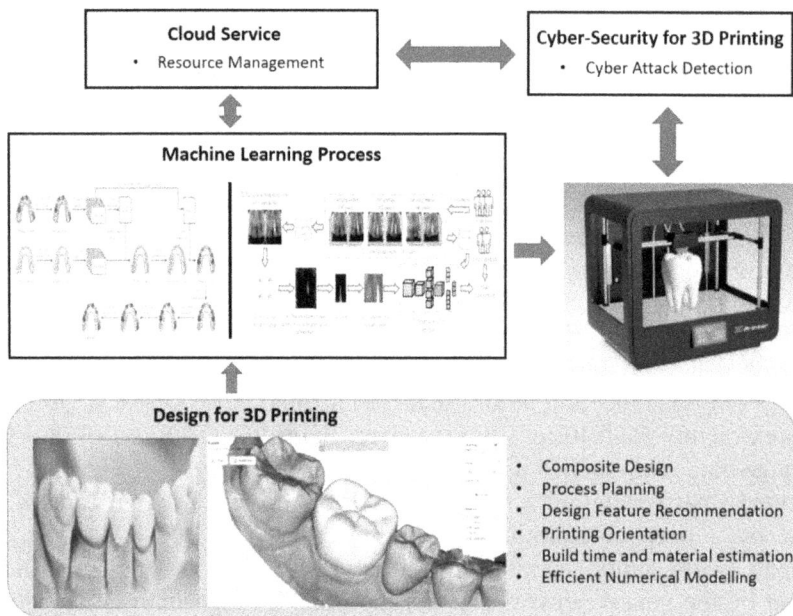

Figure 9.3. ML and CPS application in 3D printing in the dental field.

attacks and threats. When a combination of all the techniques is used, we are able to achieve the targeted model quickly.

9.8.5 ML and deep learning challenges in 3D printers

9.8.5.1 Data integration
The CPS is interconnected with different types of sensors and processing units which give data in various forms, and their integration is complex. Data without proper integration leads to numerous errors in classification.

9.8.5.2 Data shortage
The machine and deep learning algorithms train and test the data collected from the sources. The amount available for classification is significant because, the algorithm will take the important features and learn the model only from the available data. Fewer data being available for training/testing will cause the under- or over-fitting problem. This creates a critical issue in unknown data classification.

9.8.5.3 Data security
Data should be safe from cyber-attacks. The damages caused by the attacks varies according to the application domain. For example, if the data available in the 3D printer of a hospital is vulnerable to attack, it can endanger a patient's life. Hence, protection of the captured data is essential.

9.9 Conclusion

The steps involved in additive manufacturing are discussed in detail. The cyber-attack vulnerability at each stage of 3D printing is discuss. The use of cyber–physical systems and optimization algorithms for operating systems without human intervention are reported, with which the entire fabrication can become fully automated with the help of the coming firth industrial revolution. The applications of 3D printing techniques in various industries and with various process parameters are discussed in this chapter. Finally, the challenges involved in the optimization techniques for adopting additive manufacturing are reported.

References

[1] What is Industry 4.0? *IBM* https://ibm.com/in-en/topics/industry-4-0
[2] Behera D *et al* 2021 Current challenges and potential directions towards precision microscale additive manufacturing—part IV: future perspectives *Precis. Eng.* **68** 197–205
[3] Wang Y *et al* 2019 Advanced materials for additive manufacturing *IOP Conf. Ser.: Mater. Sci. Eng.* **479** 012088
[4] Xiong Y, Tang Y, Zhou Q, Ma Y and Rosen D W 2022 Intelligent additive manufacturing and design state of the art and future perspectives *Addit. Manuf.* **59** 103139
[5] Nain G, Pattanaik K K and Sharma G K 2022 Towards edge computing in intelligent manufacturing: past, present and future *J. Manuf. Syst.* **62** 588–611
[6] Mhetre G N *et al* 2022 A review on additive manufacturing technology *ECS Trans.* **107** 15355

[7] Qin J, Hu F, Liu Y, Witherell P, Wang C C L, Rosen D W, Simpson T W, Lu Y and Tang Q 2022 Research and application of machine learning for additive manufacturing *Addit. Manuf.* **52** 102691

[8] Sturm L D, Williams C B, Camelio J A, White J and Parker R 2017 Cyber-physical vulnerabilities in additive manufacturing systems: a case study attack on the STL file with human subjects *J. Manuf. Syst.* **44** 154–64

[9] Mahan T and Menold J 2020 Simulating cyber-physical systems: identifying vulnerabilities for design and manufacturing through simulated additive manufacturing environments *Addit. Manuf.* **35** 101232

[10] Faruque M O, Lee Y, Wyckoff G J and Lee C H 2023 Application of 4D printing and AI to cardiovascular devices *J. Drug Deliv. Sci. Technol.* **80** 104162

[11] De Simone V, Pasquale V D and Miranda S 2023 An overview on the use of AI/ML in manufacturing MSMEs: solved issues, limits, and challenges *Procedia Comput. Sci.* **217** 1820–9

[12] Cho S, Nam H J, Shi C, Kim C Y, Byun S H, Agno K C, Lee B C, Xiao J, Sim J Y and Jeong J W 2023 Wireless, AI-enabled wearable thermal comfort sensor for energy-efficient, human-in-the-loop control of indoor temperature *Biosens. Bioelectron.* **223** 115018

[13] Ding X, Li Z and Liu C 2022 Monolithic, 3D-printed lab-on-disc platform for multiplexed molecular detection of SARS-CoV-2 *Sens. Actuators* B **351** 130998

[14] Rahmani Dabbagh S, Ozcan O and Tasoglu S 2022 Machine learning-enabled optimization of extrusion-based 3D printing *Methods* **206** 27–40

[15] Parrilla M, Vanhooydonck A, Johns M, Watts R and De Wael K 2023 3D-printed microneedle-based potentiometric sensor for PH monitoring in skin interstitial fluid *Sens. Actuators* B **378** 133159

[16] Wang C, Tan X P, Tor S B and Lim C S 2020 Machine learning in additive manufacturing: state-of-the-art and perspectives *Addit. Manuf.* **36** 101538

[17] Tian X *et al* 2022 Roadmap for additive manufacturing: toward intellectualization and industrialization *Chin. J. Mech. Eng. Addit. Manuf. Front.* **1** 100014

[18] Mennenga M, Rogall C, Yang C-J, Wölper J, Herrmann C and Thiede S 2020 Architecture and development approach for integrated cyber-physical production-service systems (CPPSS) *Procedia CIRP* **90** 742–7

[19] Azlin M N M, Ilyas R A, Zuhri M Y M, Sapuan S M, Harussani M M, Sharma S, Nordin A H, Nurazzi N M and Afiqah A N 2022 3D printing and shaping polymers, composites, and nanocomposites: a review *Polymers* **14** 180

[20] Gao Y, Li B, Wang W, Xu W, Zhou C and Jin Z 2018 Watching and safeguarding your 3D printer: online process monitoring against cyber-physical attacks *Proc. ACM Interact. Mob. Wearable Ubiquitous Technol.* **2** 27

[21] Khanzadeh M, Rao P, Jafari-Marandi R, Smith B K, Tschopp M A and Bian L 2018 Quantifying geometric accuracy with unsupervised machine learning: using selforganizing map on fused filament fabrication additive manufacturing parts *J. Manuf. Sci. Eng.* **140** 031011

[22] Ferreira R D S B, Sabbaghi A and Huang Q 2019 Automated geometric shape deviation modeling for additive manufacturing systems via Bayesian neural networks *IEEE Trans. Autom. Sci. Eng.* **17** 584–98

[23] Shen Z, Shang X, Zhao M, Dong X, Xiong G and Wang F-Y 2019 A learning-based framework for error compensation in 3D printing *IEEE Trans. Cybern.* **49** 4042–50

[24] DeCost B L *et al* 2017 Computer vision and machine learning for autonomous characterization of AM powder feedstocks *JOM* **69** 456–65

[25] Vrabel J *et al* 2019 Classification of materials for selective laser melting by laser induced breakdown spectroscopy *Chem. Pap.* **73** 2897–905

[26] Valente R *et al* 2020 Classifying powder flow ability for cold spray additive manufacturing using machine learning *IEEE Int. Conf. on Big Data* pp 2919–28

[27] Caggiano A, Zhang J, Alfieri V, Caiazzo F, Gao R and Teti R 2019 Machine learning-based image processing for on-line defect recognition in additive manufacturing *CIRP Ann.* **68** 451–4

[28] Wang T, Kwok T-H, Zhou C and Vader S 2018 *In-situ* droplet inspection and closed-loop control system using machine learning for liquid metal jet printing *J. Manuf. Syst.* **47** 83–92

[29] Trenfield S, Awad A, Madla C, Hatton G, Firth J, Goyanes A, Gaisford S and Basit A 2019 Shaping the future: recent advances of 3D printing in drug delivery and healthcare *Expert Opin. Drug Deliv.* **16** 1–14

[30] Yu Y J 2016 Machine learning for dental image analysis arXiv: 1611.09958

IOP Publishing

3D Printed Smart Sensors and Energy Harvesting Devices
Concepts, fabrication and applications
Sanket Goel and Sohan Dudala

Chapter 10

Applications: smart sensors

Aniket Chakraborthy, Suresh Nuthalapati, H Harija, Anindya Nag and Mehment Ercan Altinsoy

This chapter highlights some of the notable work done on the application of smart sensors. There has been a prominent rise in the use of smart sensors over the years, along with an estimated rise in the near future. Since energy harvesting is one of the popular sensing applications of recent years, the fabrication and implementation of nanogenerators is highlighted. The performance of these sensors is largely dependent on the type of raw materials processed to develop them. A wide range of nanomaterials and polymers, based on their electrical conductivity, tensile strength, and biocompatibility, have been used to develop these devices. Along with their electromechanical attributes, the performances of these sensors were analyzed by using them as wearable prototypes to determine gestures and physiological movements.

10.1 Introduction

Energy is an integral component in the advancement of human civilization. One of the most significant challenges the world is currently facing is the conflict between the rising demand for energy and the increasing scarcity of available energy resources. Since the excessive use of fossil fuels has increased environmental contamination, an energy harvesting approach is used for storing energy from external environmental sources in small devices [1]. Therefore, it is critical to discover and create innovative, sustainable, and environmentally friendly forms of energy [2]. Sensor technology is integral to various applications, such as in automation, transport, food, chemical industries, and healthcare monitoring, as well as our daily activities. Promising studies have been reported on developing wearable or implantable nanogenerators that can detect human physiological signals and harvest mechanical energy from human activity [3].

Several methods have been mentioned in the literature to extract energy from our environment. Recently, sensors have been an ideal way to generate and harvest energy. Semiconducting sensors gained recognition a few decades ago [4, 5] and, as a result, silicon substrate-based sensors have gained popularity due to their distinct advantages over other sensors on the market. These silicon sensors were created using the traditional microelectromechanical (MEMS) method [6, 7], in which the electrodes are created using the photolithography technique [8, 9]. Silicon sensors have been employed in various environmental [10, 11] and industrial [12, 13] applications, but some of their shortcomings concerning biomedical applications have prompted researchers to look for alternatives. Hence, flexible sensors [14–16] have been selected as a suitable option due to their enhanced electrical, mechanical, and thermal characteristics. The flexible sensors have been created using various nanomaterials [17–19] and polymers [20–22] for different applications. Some of the common nanomaterials used are carbon allotropes [23–25], metallic nanowires [26–28], and nanoparticles [29–31]. These sensors were created using a roll-to-roll manufacturing process and various printing methods. Common fabrication methods include screen printing [32, 33], inkjet printing [34, 35], laser ablation [36, 37], and 3D printing [38, 39]. These sensors have been capable of generating and harvesting considerable amounts of energy. There are two types of energy harvesters, namely piezoelectric nanogenerators (PENGs) [40, 41] and triboelectric nanogenerators (TENGs) [42, 43]. Both these types of nanogenerators have been preferred over traditional electromagnetic generators due to their simpler structure and higher output power density [44]. These two energy harvester types have been the subject of ongoing research. However, TENGs have proven far superior to PENGs in terms of high output power, a wider variety of materials for forming the sensors, a simpler fabrication technique, lower cost, and less weight [45].

Recently, various applications have used these nanogenerators as intelligent sensors for energy harvesting. According to the market study shown in figures 10.1(a) and (b) [46, 47], using smart sensors will increase revenue from 45.8 billion USD to 104.5 billion USD, growing at a compound annual growth rate of 17.9% between the years 2022 and 2027. Smart sensor usage has decreased in recent years due to the COVID-19 pandemic, but there has been a discernible rise from the pre- to the post-pandemic era.

10.2 Types of smart sensors

Different kinds of sensors have been developed at the academic and industrial levels and subsequently used as smart sensing prototypes. Some of the significant ones have been highlighted here. One of the initial works [48] highlights the development of smart-textile-based TENGs for energy harvesting. An easy and cost-effective method was used to fabricate poly(3,4-ethylene-dioxythiophene) polystyrene sulfonate (PEDOT:PSS) functionalized sensors. An arch-shaped sensor was designed and attached to the finger joints to understand sign language and control robot fingers. It was found that from the characterization analysis, the energy harvester fabricated by coating the PEDOT:PSS solution provided better results. The textile's various body motions, including simple hand tapping and foot stepping, were examined for potential energy harvesting. The PTFE film produces negative charges after serving

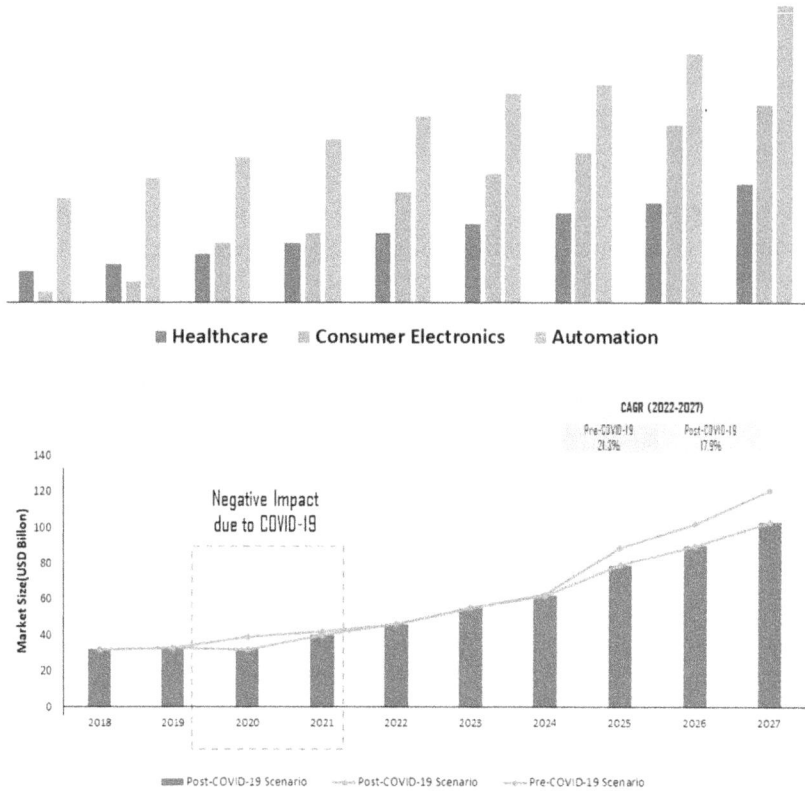

Figure 10.1. (a) Global smart sensor market by end users (data from [46]). (b) Estimation of the rise in the use of smart sensors between 2018 and 2027 (data from [47]).

as a positive triboelectric layer and electrode with the functionalized textile. The output voltage was measured with a 100 M probe during stepping with the feet and tapping with the hands. Experiments have been carried out to determine the matching power curve for the textile-based TENG under various loads for tapping with the hands and feet. The output voltage was captured by repeatedly moving the hand at a 2 Hz frequency. Hand tapping with 2.67 mW output power, measured with a load resistance of 9 MΩ, could have produced an average of 460 V. The same frequency of 2 Hz was also used for foot tapping, which produced an output voltage of 540 V and a maximum power of 3.26 mW with 14 MΩ of resistance. The output voltage of the TENG remained constant even after 4 h of continuous contact separation motion, demonstrating the durability of the functionalized textile. The textile's ability to be washed was also looked into, and after each wash, the output voltage was measured. The output voltage remained at 81% of its initial value after the textile was washed several times, another cloth was used to rub it, and the voltage was measured.

To harvest mechanical energy, cement nanocomposites with improved dielectric properties have also been used to create a cement-based TENG [49]. Since titanium oxide (TiO_2) is non-toxic, chemically inert, has high permittivity, and is easy to disperse in cement, it is one of the most notable fillers used in this method. Due to its self-cleaning and air pollution treatment photocatalytic qualities, it is widely used in cement. For the TENG, a cement–TiO_2 composite was developed in this work. The effect of the TiO_2 fraction in cement was studied in terms of energy conversion effectiveness and dielectric properties. It was discovered that 0.2 wt% of TiO_2 can improve the energy conversion of the composite, resulting in a power density of up to 265 mW m^{-2}.

Another study [50] demonstrated the fabrication and utilization of flexible PAA-RGO-PANI (polyaniline, reduced graphene oxide, and polyacrylic acid hydrogel). The purpose of the developed sensor was to be implemented in smart farm systems. Three layers of polyimide film (PI), a freeze-dried PAA-RGO-PANI hydrogel, and an aluminum film piece were sandwiched together to create a PRP-TENG clean energy harvesting device. The aluminum film served the purpose of the current collector in the above sensor. The electrical performance of various sample compositions was assessed in key areas, including the effectiveness of the nano-generator. The samples with RGO nanofiller had better electrical performance than those without. The concentration of the RGO nanofiller was increased from 5 to 15 mg ml^{-1}, which had a significant positive impact on electrical performance. Additionally, when PANI was applied to the PAA-RGO samples, electric perform-ance significantly improved as RGO and PANI were evenly coated to obtain the porous PAA-RGO-PANI nanocomposite.

Due to the extremely high specific surface area and ultra-low density produced, it helps to maintain a significant amount of nanoscale conductive fillers (RGO). This enables extensive contact area when triboelectrification occurs and provides numer-ous electrons on the nanocomposite surface. To measure the PRP-TENG's electrical output and gauge the device's ability to capture acoustic energy, a speaker was used as an acoustic signal. The performance of the PRP-TENG in this study of acoustic energy harvesting makes use of an acoustic source at 85 dB and 1 kHz. The findings demonstrated that the PAA-RGO-PANI nanocomposite could produce a higher output current and voltage when additional RGO was present. Stable energy harvesting from an acoustic source by the PRP-TENG was obtained, which has 15 mg ml^{-1} RGO. When a steady acoustic energy source was delivered at 85 dB and 1 kHz continuously for 50 s, the PRP-TENG with 15 mg ml^{-1} RGO could maintain a V_{oc} of 12.67–1.13 V and a stable I_{sc} of 15.67–6.89 A.

Disposable smart microfluidic platform microchips (DIS-μChip) were also fabricated and utilized for multifunctional applications [51]. The chip is fully automated for pressure control and monitoring inlet and outlet flow. A polydime-thylsiloxane (PDMS) and polyethylene terephthalate (PET) superstructure was fabricated. Functional components in the substrate enable the generation of an energy field that can pass through the PET thin film, regulate the fluid in the substrate's microchannels, and serve as sense electrodes for flow sensors integrated into the chip. The silicone nanoparticles were coated on one side of a PET thin film. Because of the inexpensive components and straightforward fabrication, the super-strate could be used in a disposable manner to reduce biocontamination and increase patient safety in biological, therapeutic, and pharmaceutical applications. Since the

PET thin film serves as the superstrate's microchannel's bottom layer, it is possible to create multiple flow sensors and combine them on-chip with microfluidic features for the disposable film-chip technology. The chip allows for an ultra-low flow of less than 100 nl in volume due to its simple experimental set-up and tubing, making automatic fluidic control on the chip possible. The detectable full-scale flow rate range can be configured according to the applications. The DIS-μChip can be used in various applications by altering the microchannel structure.

Inspired by wearable sensors, Zhao et al [52] developed a smart, plant-based sensor that can be used as a pesticide on crop surfaces. An electrochemical-based wearable biosensor for plants can be placed on a plant surface to detect organophosphorus pesticides. Direct writing was used to create 3D porous laser-induced graphene on a PI film. The pattern was then covered with PDMS and heated for 120 min at 75 °C. Later, to accommodate for uneven crop surfaces and likely environmental deformation, a stretchable and flexible plant-wearable sensor was created. Then, it was altered with organophosphorus hydrolase and given a biocompatible semisolid electrolyte so that the wearable biosensor on the plant could specifically capture and recognize the methyl parathion on its surface. A portable electrochemical workstation was employed to study the electrochemical behavior.

Wang et al [53] described a flexible TENG with a magnetic field multifunctional energy harvester. In daily life, human hands are essential tools. Numerous tasks are completed with only our hands because of how skilfully the finger joints have evolved. They served as the inspiration for the creation of a TENG-based glove for energy collection and monitoring of external stimuli. the TENG was assembled on a hand mold, and eco-flex/CI was poured over and dried to connect the array. The glove was comfortable to wear on the hands and flexible enough to adjust to intricate mechanical deformations. The TENG detected the bending excitation, and the maximum output voltage generated at the center of the TENG was 0.76 V. The back of the hand had an array of 3 × 3 units that could recognize and display various compression forces. The highest voltage delivered by the center TENG was 0.76 V, showing that it can sustain a larger compression force. The pressure detection while holding a ring box and an apple and the voltage changes could be observed. The voltages were minimal when the middle and ring fingers were bent. However, the signals became stronger after a solid box was grasped, and the compression forces need to be measured at the finger joints. 1.45 V was the maximum voltage that was measured by the force of the thumb while grasping the box. The array's self-powered, high sensitivity, and real-time monitoring characteristics allow use in wearable human healthcare, smart robots, and safety alarm systems.

Shuai et al [54] developed stretchable and self-healing hydrogen fibers using a constant dry and wet spinning approach. Poly (NAGA-co-AAm) hydrogel fibers were obtained by converting physically cross-linked hydrogel precursors and polymerizing them with acrylamide (AAm) and N-acryloyl glycine amide (NAGA). The PNA hydrogel fibers achieved a high textile strength of 2.27 MPa. They were coated with elastomeric poly (methyl acrylate) (PMA). The PNA/PMA core–sheath fiber exhibited remarkable resistance to water absorption and evaporation after being formed. The PNA/PMA fibers were demonstrated for strain-sensing applications. The TENG textiles were weaved from the PNA/PMA fibers. The fabric was soft and had good stretchability. The dielectric PMA sheath was the triboelectrification layer, while the conductive core was the electrode. The PMA became positively charged when polytetrafluoroethylene (PTFE) was chosen as the counter material. If they became separated, opposite charges were induced on the conductive hydrogel fibers as the PNA/PMA interfaces have to balance the charges. The current flow stopped when the PTFE was maintained at a certain distance and all its static charges were screened. Likewise, the charges flow in other directions when the PTFE comes in contact with the TENG fabric, creating an opposing current in the circuit. The measured output voltage of the TENG textile was -36 V a 1.25 Hz in a contact separation motion. The TENG textile's output was measured at the extended state, and the output voltage moderately decreases while the textile was stretched from a range of 0%–100%. The output voltage was recovered when the strain was released. To demonstrate the TENG textile as a power source, the AC output was turned to DC and was stored in a capacitor (6 μF). The capacitor was able to charge to 3.5 V in 260 s and could supply 14 s of power to an electronic device.

Marra *et al* demonstrated a novel method for creating piezoresistive fabrics based on water-based ink containing graphene nanoparticles (GNPs) [32]. A screen-printing technique was employed in which commercial ink could be adequately loaded with graphene nanoplatelets to produce sensor fabrics. The sensor was developed and measured in a controlled laboratory environment with a relative humidity (RH) of 40% and a temperature of 23 °C. Quasi-static mechanical tests were conducted on the screen-printed fabric; when compared to the raw fabric, the material's stiffness increased after the coating procedure, depending on the filler wt%. The electromechanical tests, where piezoresistive response was evaluated on the coated fabric, revealed an increase in sensitivity with strain at maximum strain (5%), and even after washing, the gauge factor (GF) was 30, indicating strong repeatability during work cycles. Resistance versus time for the sample contains GNPs of 3.0 wt% before washing and after washing with water. A piezoresistive material that is very sensitive and has repeatable behavior was integrated into a synthetic fabric. A variety of deformations were explored, and the sensor has a potential application for use in wearable electronics focusing on monitoring various biometric parameters such as breathing or heart rate.

Veeralingam *et al* [55] demonstrate that a wearable breath sensor with outstanding switching and response can be developed using a FeS_2-based memristor device fabricated in a single hydrothermal step with a cost-effective, biodegradable, flexible cellulose paper substrate. With a voltage, i.e. $+0.25$ and -0.5 V, and very low power consumption in the range of a few microns to nanowatts, the device demonstrates an outstanding bipolar resistive switching mechanism. The breath sensor was kept away from the mouth maintaining a distance of 5 cm and allowing the person to breathe into the device (figure 10.2(a)). When exposed to exhaled breath, it was discovered that the sensor's resistance decreased. The performance of the developed breath sensor was further tested under various exhalation situations, including nasal and oral breath. The sensor was also placed at a distance of 10 cm

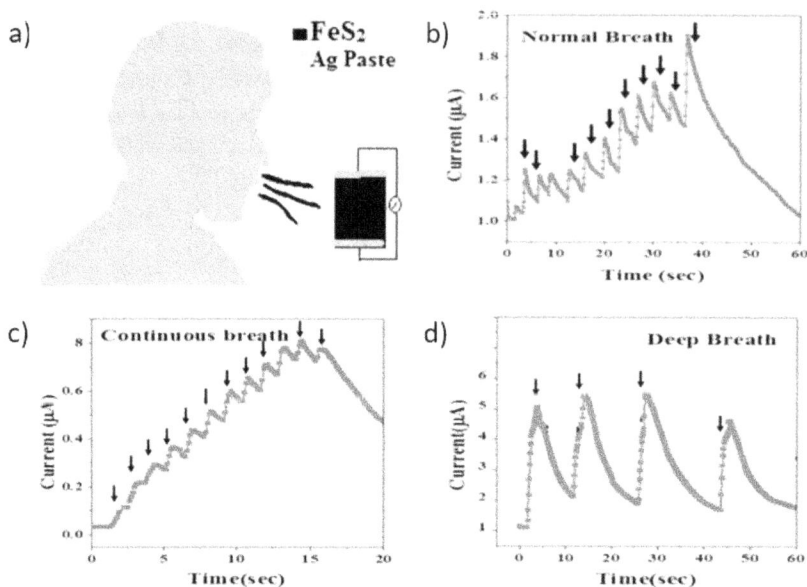

Figure 10.2. (a) Schematic pictorial representation of a breath sensor. (b) Normal breathing condition, (c) continuous breathing condition, and (d) deep breathing condition. (Reproduced with permission from [55]. Copyright 2020 Elsevier.)

from the nose, and the response of the nasal breath was recorded for 60 s. The breath rate was manually counted for 60 s, and it was found that it is 14 breaths per second, consistent with how people normally breathe. When the sensor was placed 10 cm from the mouth and the subject advised to breathe for 60 s, it was found that the current flow through the sensor increased when there was nasal breath, as shown in figure 10.2(b). The observation could be explained by more water molecules in the exhaled oral breath, increasing conductivity, and the nasal mucosa decreasing airflow through the breath sensor by filtering it. Additionally, the sensor responded differently to continuous and deep breathing, as shown in figures 10.2(c) and (d). The breath sensor was able to detect breathing rate during dynamic motions such as fast walking, jumping, and running to obtain the sensor's responses to normal, continuous, and deep breaths.

Reddy *et al* [56] developed a multifunctional sensor based on p-NiO/n-CdS on indium tin oxide (ITO) coated polyethylene terephthalate (PET) as a flexible substrate. The sensor could be used for breath monitoring by using a solution-processed deposition technique at low temperatures. The manufactured p-NiO/n-CdS device could recognize several breath patterns, showing that the device has breath-sensing capacity. An individual's (healthy human) breathing pattern was recorded, and the sensor was exposed to inhalation and exhalation to show the NiO/CdS device's applicability. A decrease in sensor current was seen during inhalation, and a sudden increase in current was noted after exhalation. Similar results were also obtained from the above work. It shows that the water molecules slowly desorb from the NiO layer's surface, restoring the current to its initial value. Several breathing patterns, such as rapid and moderate breath patterns, were tried on the device. A 30% drop in the breath pattern is seen with slow breathing. Similarly, the sensor was evaluated for a fast-breathing pattern, where there was a change in the breath rate of about 55% over normal breathing. The process was repeated seven times after allowing the sensor current to drop to a level close to the baseline. The recorded breath pattern features show that the device can identify breath patterns that were crucial for breath sensors. The sensor was also examined for flexibility and longer stability. It was noted that there had been barely any change in the device's functioning, indicating its reliability.

Li *et al* [57] describe a self-powered, miniature aluminum sensor to measure breath humidity. An interdigitated capacitor sensor is designed on a thin film of sputtered aluminum and silicon (Al 1% Si). As previously mentioned, the thermo-plastic urethane (TPU) along with a electro-spin fibrous network TPUN/c-multi-walled carbon nanotube (MWCNT) pressure sensor can be used for the full range of human motion sensing as e-skin or wearable devices due to its outstanding adaptability, high sensitivity, durability, and wide pressure range. It shows measured resistance signals of finger pressing pressure at various frequencies. As the sensor was flexible, it could fold into a 'U' shape, as shown in figure 10.3(a). A low R/R_0 value results from the enhanced contact points of the process between the conductive material and interdigitated electrode. It was noted that the sensor returns to its flat shape after being bent several times, ensuring the stability of wearable technology. A plaster was used to attach the sensor to human skin, aiming to monitor physiological signals better (figure 10.3(b)). Large and delicate motions were the categories describing the sensor's application scenarios. The sensor was attached to a finger to track real-time finger-bending motions. A regular and stable resistance response

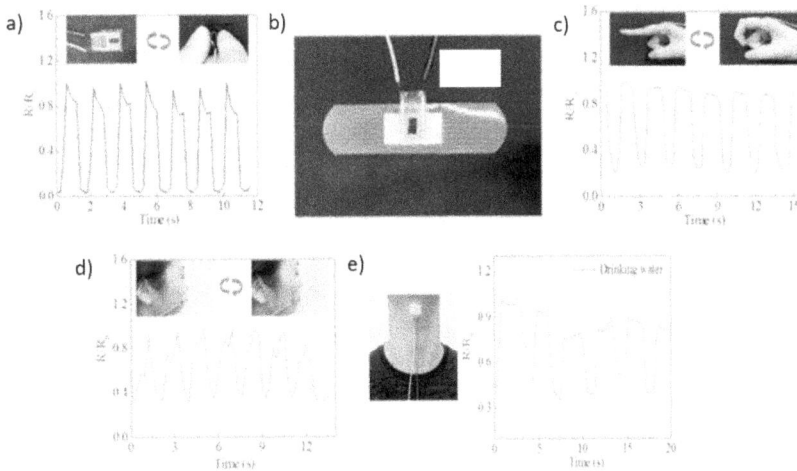

Figure 10.3. (a) Sensor bending in a U shape, (b) image of a pressure sensor attached to a plaster, (c) finger joints bending, (d) cheek bulging, and (e) drinking water. (Reproduced with permission from [57]. Copyright 2021 Elsevier.)

was obtained by repeated finger bending at a bend angle of around 60° (figure 10.3(c)). Some equivalent measurements were made with continuous bulging and swallowing water when the sensor was placed on the cheek and neck, as shown in figures 10.3(d) and (e). According to the sensory receptivity to cyclic motions, the sensor exhibits outstanding stability and reproducibility during the monitoring process.

Li *et al* [58] suggested preparing MWCNT/PDMS microspheres (MSs) using the oil-in-water Pickering emulsion process to create a 3D elastic conductive network. The strain sensor's two ends were connected to a stepper motor to measure the sensor's capacity to detect strain. The electrical response of the sensor was measured using a digital source meter. A tensile stress was applied to the sensor at 0.5 m m s^{-1}. It was observed that the relative change in resistance increased as the elongation varied from 0% to 30%, but the strain sensor had a larger strain limit of over 40%. The device showed excellent sensitivity and obtained a GF of around 7.22. It also had a high conductivity of $157.64 \ \Omega$ cm. The movement of the finger knuckles was observed when two ends of the device were attached to the finger. The knuckle stretched when the finger was bent, which reduced the conductive channel and increased resistance. The sensor exhibited distinct electrical signals for different gestures, such as 'OK', 'victory', and 'great'.

Wang *et al* [59] reported a self-powered triboelectric tactile sensor that was fabricated with a PDMS triboelectric layer having a multi-level structure and a composite electrode made of PDMS and eutectic gallium-indium (EGaIn) alloy. These triboelectric tactile sensors developed with EGaIn/PDMS composite electrodes can sustain high tactile forces while maintaining good electrical conductivity. Additionally, these tactile sensors can respond to external pressures even if they are damaged. A controlled puncturing experiment was used to study the triboelectric tactile sensor's pressure response. The damaged sensor performed a stable response for various applied pressures and generated constant output voltages when an external pressure of 872.6 Pa was applied. The damaged triboelectric tactile sensor was like the original one and reported two sensitivities ($K_1' = 0.293$ V Pa^{-1} and $K_2' = 0.087$ V Pa^{-1}) in two separate pressure ranges. The sensitivities of the sensor under severe damage conditions are close to the values of the undamaged sensor, and the value of K_1' was the same as K_1, demonstrating the exceptionally stable operation of this tactile sensor under extreme mechanical damage.

Wu *et al* [60] reported CNT and MXene thermoplastic polyurethane (CNTs/ MXene-TPU) based hybrid fibers using the wet spinning method, in which the major structural component was TPU molecular chains. The conductive networks were made up of CNTs and MXene. Pure TPU fiber has a tensile strength of 28 MPa and a more than 300% break elongation. The tensile strength and maximum elongation decreased steadily as the number of conductive fillers increased. The strain and strength at breaking were estimated as 100% and 12 MPa, respectively, when the composite of CNTs and MXene increased up to 20 wt%. The change in resistance increased with an increase of conductive fillers at a constant tensile strain of 50%. The fiber's mechanical qualities decrease as the filling increases, but the resistance's sensitivity increases. These hybrid fibers exhibited higher electrochemical performance due to the fiber's porosity structure, and its volumetric specific capacitance can reach as high as 8.8 F cm^{-3}. As a result, these hybrid fiber-based supercapacitors can be stretched and exhibited an energy density of 1.16 mWh cm^{-3}.

Almukhlifi et al [61] introduced an effective technique for developing flexible sensors using PEDOT-PSS conductive composites. These composites were synthesized with DMSO, PVA, and SnO$_2$ nanoparticles. The composite film's response was studied for various humidities ranging from 0%–100% RH to explore the impact of different humidity conditions on liquid petroleum gas (LPG) sensing performance. At LPG concentration of 100 ppm, the film's LPG-detecting response was tested under various humidity conditions. The composite film's LPG sensing response does not exhibit a linear dependence on humidity; rather, its response was generally stable with a slight variance (10%) for humidities in the range of 0%–100% RH. The addition of PVA and DMSO to the polymer matrix increased the cohesion between the molecules and enhanced the stability of the sensor toward moisture, which was attributed to the stability of the composite performance under various moisture circumstances. At 40% RH, the sensor response was shown to perform best, with additional reductions at higher concentrations of % RH. The sensor composite was extremely stable for high humidity conditions, as evidenced by the minimal variation in gas response in the humidity range of 0%–100% RH. The composite film's variation in conductivity and sensitivity toward various angles between 0° and 180° was measured. It was observed that the composite's conductivity maintained its stability over bending angles, which is interesting since it shows how well the flexible composite performed electrically when subjected to mechanical deformations. The composite film's sensitivity at 100 ppm for various bending angles (0°–180°) shows a slight decrease from 79% to 76.7% (a less than 3% reduction). A smaller exposure region for LPG gas molecules at different bending angles may be responsible for the slight decrease in sensitivity. This suggests the composite film has superior sensing properties during mechanical deformations such as bending.

Xue et al [62] reported that flexible strain sensors could be manufactured using a liquid power-ultrasound system. MWCNT nanomaterials were deposited onto a PDMS substrate having a thickness of 200 μm using this technique, which uses acoustic cavitation and acoustic streaming in a water solution sonicated by power ultrasound at 19.9 kHz. The constructed strain sensor exhibits high sensitivity (GF = 8.468.3), a broad sensing area (480% strain), and exceptional stability (>10 000 cycles at 50% strain). To demonstrate the potential uses of the manufactured flexible strain sensors, the sensor was attached to various human body parts and other objects. Then, sensing reactions to the motions were recorded. To track the degree of bending in real time, a latex glove was used on a finger. As shown by the measured response curve, the response curve is capable of accurately reflecting the degree of finger bending. To monitor wrist bending, the sensor was attached to a wrist guard. The wrist's bending and recovery could be detected and tracked with excellent repeatability. The sensor was fixed to a human neck to demonstrate how well it could detect the minute muscle movement that occur during swallowing. The sensor can view the swallowing procedure in real time. The sensor was attached to a stocking to identify the motions of knee flexion, maintaining time, and sit-up/ standing. This demonstrated the sensor's potential applications in medical training, at-home rehabilitation, and sports monitoring. The pulse is a crucial physiological indicator of heart rate and it is essential for the flexible strain sensors to be able to recognize pulse signals.

To test the practicality of commercial application, Hossain *et al* [63] fabricated hundreds of meters of textile materials based on a pressure-sensitive semiconductor. Stainless steel yarn and a cross-stitching approach were used, and a conductive textile surface was created for the top and bottom electrodes. A three-layer structure was used for the piezoresistive sensor technology. The middle layer was made of a semi-conductive substance and was surrounded by conductive top and bottom layers. In this case, the characteristics of the semiconductor material were crucial for enhancing the high precision and steadiness of the resistive response to pressure. The characterization of the smart-textile sensor in several areas, including weight recognition, product movement or replacement, linearity, repeatability, etc, has proven its validity. The resistance deviates when a product with a different weight is kept on the sensor. The weight causes a fine edge, representing a shift in values of around 22, 27, and 29 for the number of goods (1, 2, and 3), weighing 500 g. Compared to products that have been added to the shelf with greater weights, the product with the lowest weight (500 g) tends to have a higher variation in the resistance. It is caused by the sensor's high sensitivity to changes in resistance at low product weights, which stabilizes with small changes in resistance at larger product weights up to 1500 g.

Shen *et al* [64] showed that different tensile states could detect the sensor's current–voltage characteristic curve. A strong linear relationship between the voltage and current revealed the sensor's good ohmic performance even when not under strain. The slope of the $I–V$ curve gradually decreased as the strain increased, demonstrating that the corresponding resistance increases as the strain does. Increment step sizes were increased by 5% every 5 s, and the strain applied to the sensor ranged from 0% to 50%. The interaction between the initially used MWCNTs changed as the strain increased from 0% to 50%, the corresponding resistance increased progressively, and the entire phase appeared to be growing in steps. The ability of the sensor to detect and distinguish between distinct motions while simultaneously monitoring its relative resistance to various strains in real time was demonstrated by the test.

Tian *et al* [65] revealed that piezoresistive sensors continue to have issues with high initial costs, a weak link between conductive materials and substrates, a restricted sensing area, and bad ergonomics. A 3D piezoresistive sensor composed of tightly compressed bicomponent fibers on a nonwoven substrate was produced. Due to the distinct core–sheath structure of the fibers and a coating method that uses thermal assistance to make the nonwoven material conductive, MWCNTs adhere quickly and firmly to the fiber surface. The sensor demonstrated that it had a broad sensing range from 0 to 131.32 kPa, a response time of 105 ms, and a relaxation time of 156 ms, exceptional flexibility, and long-term reliability. These characteristics are based on the multi-layered fibrous structure, because of which the sensor provided a high sensitivity of 5.57% and 0.113% kPa^{-1} in the ranges of 0–7 and 30–131.32 kPa, respectively.

The work done in [66] shows humidity and piezoresistive force sensors developed with graphene-coated cellulose paper (CGN) using vacuum filtration (V-CGN) and dip-coating (D-CGN). Both sensors' piezoresistive performances were evaluated under different applied loads. Initially, the sensor's resistance comprised various types, including tunneling resistance, intrinsic electrode resistance, and resistance resulting from the contacts between the electrodes and graphene. The piezoresistive performance of a sensor built on a V-CGN was tested under various loads. After applying the load, the sensor's resistance decreased. The sensor resistance is sharply decreased when more pressure is placed on the sensor. To illustrate the sensor's potential and favorable features, it was tested using finger tapping. The sensor was sensitive to finger touch and responded quickly. The results demonstrated the sensor's sensitivity, durability, and repeatability, which may be advantageous for applications such as thin film sensors. The responses of both sensors to various loads were determined. The performance of the D-CGN sensor was observed while finger tapping a 1 N applied load. The D-CGN sensor's performance when the resistance varies as a result of tapping the finger was determined. The response of the D-CGN sensor was similar to that of the V-CGN sensor. Change is resistance when the sensor holds and releases the load. The sensor's resistance change was less than that for the V-sign. The sensors' response times were the same (1 s), but the D-CGN sensor's recovery time was higher (5–6 s).

10.3 Conclusion

The chapter showcases some of the significant work done on the development of nanogenerator and other types of sensors and their use for specific applications. Different kinds of fabrication techniques have been utilized to process the nano-materials and polymers in forming the sensors. Due to the mechanical flexibility of these sensors, they have been used for strain-induced tactile and energy harvesting applications. These sensors are capable of operating in smart sensing with high repeatability, excellent reproducibility, wide operating range, and high sensitivity. As a future work, these sensors need to be integrated with various communication protocols and signal-conditioning circuits to utilize the sensing systems in real-time applications. The combination of different nanomaterials should be encouraged to strengthen the interfacial bonding with the chosen polymeric substrates. The toxicity of the processed materials with respect to their human interaction should be studied further to increase the possibility of developing point-of-use devices for different applications.

References

[1] Maruccio C, Montegiglio P and Kefal A 2020 Parameter identification strategy for online detection of faults in smart structures for energy harvesting and sensing *Procedia Struct. Integr.* **28** 2104–9

[2] Zou Y, Bo L and Li Z 2021 Recent progress in human body energy harvesting for smart bioelectronic systems *Fundam. Res.* **1** 364–82

[3] Hu S, Shi Z, Zhao W, Wang L and Yang G 2019 Multifunctional piezoelectric elastomer composites for smart biomedical or wearable electronics *Composites* B **160** 595–604

[4] Moseley P T 2017 Progress in the development of semiconducting metal oxide gas sensors: a review *Meas. Sci. Technol.* **28** 082001

[5] Sze S M 1994 *Semiconductor Sensors* (New York: Wiley)

[6] Tilli M, Paulasto-Krockel M, Petzold M, Theuss H, Motooka T and Lindroos V 2020 *Handbook of Silicon-based MEMS Materials and Technologies* (Amsterdam: Elsevier)

[7] Bhatt G, Manoharan K, Chauhan P S and Bhattacharya S 2019 MEMS sensors for automotive applications: a review *Sensors for Automotive and Aerospace Applications. Energy, Environment, and Sustainability* ed S Bhattacharya, A Agarwal, O Prakash and S Singh (Singapore: Springer) pp 223 39

[8] Nag A, Zia A I, Mukhopadhyay S and Kosel J 2015 Performance enhancement of electronic sensors through mask-less lithography *9th Int. Conf. on Sensing Technology* (Piscataway, NJ: IEEE) pp 374–9

[9] Nag A, Zia A I, Li X, Mukhopadhyay S C and Kosel J 2015 Novel sensing approach for LPG leakage detection: part I—operating mechanism and preliminary results *IEEE Sens. J.* **16** 996–1003

[10] Alahi M E E, Nag A, Mukhopadhyay S C and Burkitt L 2018 A temperature-compensated graphene sensor for nitrate monitoring in a real-time application *Sens. Actuators* A **269** 79–90

[11] Alahi M E E, Xie L, Mukhopadhyay S and Burkitt L 2017 A temperature-compensated smart nitrate-sensor for the agricultural industry *IEEE Trans. Ind. Electron.* **64** 7333–41

[12] Nag A, Zia A I, Li X, Mukhopadhyay S C and Kosel J 2015 Novel sensing approach for LPG leakage detection—part II: effects of particle size, composition, and coating layer thickness *IEEE Sens. J.* **16** 1088–94

[13] Barhoumi L *et al* 2017 Silicon nitride capacitive chemical sensor for phosphate ion detection based on copper phthalocyanine–acrylate polymer *Electroanalysis* **29** 1586–95

[14] Mukhopadhyay S C, Suryadevara N K and Nag A 2021 Wearable sensors and systems in the IoT *Sensors* **21** 7880

[15] Afsarimanesh N, Nag A, Sarkar S, Sabet G S, Han T and Mukhopadhyay S C 2020 A review on the fabrication, characterization, and implementation of wearable strain sensors *Sens. Actuators* A **315** 112355

[16] Nag A, Mukhopadhyay S C and Kosel J 2017 Wearable flexible sensors: a review *IEEE Sens. J.* **17** 3949–60

[17] Nag A, Nuthalapati S and Mukhopadhyay S C 2022 Carbon fiber/polymer-based composites for wearable sensors: a review *IEEE Sens. J.* **22** 10235–45

[18] Han T, Nag A, Mukhopadhyay S C and Xu Y 2019 Carbon nanotubes and its gas-sensing applications: a review *Sens. Actuators* A **291** 107–43

[19] Nag A, Mitra A and Mukhopadhyay S C 2018 Graphene and its sensor-based applications: a review *Sens. Actuators* A **270** 177–94

[20] Yuan C, Tony A, Yin R, Wang K and Zhang W 2021 Tactile and thermal sensors built from carbon–polymer nanocomposites—a critical review *Sensors* **21** 1234

[21] Spychalska K, Zając D, Baluta S, Halicka K and Cabaj J 2020 Functional polymers structures for (bio) sensing application—a review *Polymers* **12** 1154

[22] Sappati K K and Bhadra S 2018 Piezoelectric polymer and paper substrates: a review *Sensors* **18** 3605

[23] Villarreal C C, Pham T, Ramnani P and Mulchandani A 2017 Carbon allotropes as sensors for environmental monitoring *Curr. Opin. Electrochem.* **3** 106–13

[24] Gao J, He S, Nag A and Wong J W C 2021 A review of the use of carbon nanotubes and graphene-based sensors for the detection of aflatoxin M1 compounds in milk *Sensors* **21** 3602

[25] Nag A, Alahi M, Eshrat E, Mukhopadhyay S C and Liu Z 2021 Multi-walled carbon nanotubes-based sensors for strain sensing applications *Sensors* **21** 1261

[26] Liu L, Jiang Y, Jiang J, Zhou J, Xu Z and Li Y 2021 Flexible and transparent silver nanowires integrated with a graphene layer-doping PEDOT: PSS film for detection of hydrogen sulfide *ACS Appl. Electron. Mater.* **3** 4579–86

[27] Rafiee Z, Roshan H and Sheikhi M H 2021 Low concentration ethanol sensor based on graphene/ZnO nanowires *Ceram. Int.* **47** 5311–7

[28] Zhang S *et al* 2020 A flexible bifunctional sensor based on porous copper nanowire@ IonGel composite films for high-resolution stress/deformation detection *J. Mater. Chem.* C **8** 4081–92

[29] Yoon S and Kim H-K 2020 Cost-effective stretchable Ag nanoparticles electrodes fabrication by screen printing for wearable strain sensors *Surf. Coat. Technol.* **384** 125308

[30] Zhang Y *et al* 2020 A flexible non-enzymatic glucose sensor based on copper nanoparticles anchored on laser-induced graphene *Carbon* **156** 506–13

[31] Khalifa Z, Zahran M, Zahran M A and Azzem M A 2020 Mucilage-capped silver nanoparticles for glucose electrochemical sensing and fuel cell applications *RSC Adv.* **10** 37675–82

[32] Marra F, Minutillo S, Tamburrano A and Sarto M S 2021 Production and characterization of graphene nanoplatelet-based ink for smart textile strain sensors via screen printing technique *Mater. Des.* **198** 109306

[33] Albrecht A, Salmeron J F, Becerer M, Lugli P and Rivadeneyra A 2019 Screen-printed chipless wireless temperature sensor *IEEE Sens. J.* **19** 12011–5

[34] Yi Y, Ali S and Wang B 2019 An inkjet-printed strain sensor with a carbon-silverpolyimide topology *IEEE Int. Conf. on Flexible and Printable Sensors and Systems* (Piscataway, NJ: IEEE) pp 1–3

[35] Nayak L, Mohanty S, Nayak S K and Ramadoss A 2019 A review on inkjet printing of nanoparticle inks for flexible electronics *J. Mater. Chem.* C **7** 8771–95

[36] Nag A, Mukhopadhyay S C and Kosel J 2017 Sensing system for salinity testing using laser-induced graphene sensors *Sens. Actuators* A **264** 107–16

[37] Nag A and Mukhopadhyay S C 2018 Fabrication and implementation of printed sensors for taste sensing applications *Sens. Actuators* A **269** 53–61

[38] He S, Feng S, Nag A, Afsarimanesh N, Han T and Mukhopadhyay S C 2020 Recent progress in 3D printed mold-based sensors *Sensors* **20** 703

[39] Han T, Kundu S, Nag A and Xu Y 2019 3D printed sensors for biomedical applications: a review *Sensors* **19** 1706

[40] Ye S *et al* 2019 High-performance piezoelectric nanogenerator based on microstructured P (VDF-TrFE)/BNNTs composite for energy harvesting and radiation protection in space *Nano Energy* **60** 701–14

[41] Yuan H, Lei T, Qin Y and Yang R 2019 Flexible electronic skins based on piezoelectric nanogenerators and piezotronics *Nano Energy* **59** 84–90

[42] Luo J and Wang Z L 2020 Recent progress of triboelectric nanogenerators: from fundamental theory to practical applications *EcoMat* **2** e12059

[43] Hwang H J, Yeon J S, Jung Y, Park H S and Choi D 2020 Extremely foldable and highly porous reduced graphene oxide films for shape-adaptive triboelectric nanogenerators *Small* **17** 1903089

[44] Zhao J *et al* 2019 Remarkable merits of triboelectric nanogenerator than electromagnetic generator for harvesting small-amplitude mechanical energy *Nano Energy* **61** 111–8

[45] Rathore S, Sharma S, Swain B P and Ghadai R K 2018 A critical review on triboelectric nanogenerator *IOP Conf. Ser.: Mater. Sci. Eng.* **377** 012186

[46] Global smart sensor market by end users *Markets and Markets* https://marketsandmarkets. com/Market-Reports/smart-sensor-market-43119772.html

[47] Estimation in the use of smart sensors for 2018–2027 *Maximize Market Research* https:// maximizemarketresearch.com/market report/smart sensor market/127244/

[48] He T *et al* 2019 Beyond energy harvesting-multi-functional triboelectric nanosensors on a textile *Nano Energy* **57** 338–52

[49] Sintusiri J, Harnchana V, Amornkitbamrung V, Wongsa A and Chindaprasirt P 2020 Portland cement–TiO_2 triboelectric nanogenerator for robust large-scale mechanical energy harvesting and instantaneous motion sensor applications *Nano Energy* **74** 104802

[50] Hsu H H *et al* 2021 Self-powered and plant-wearable hydrogel as LED power supply and sensor for promoting and monitoring plant growth in smart farming *Chem. Eng. J.* **422** 129499

[51] Kim J, Cho H, Kim J, Park J S and Han K-H 2021 A disposable smart microfluidic platform integrated with on-chip flow sensors *Biosens. Bioelectron.* **176** 112897

[52] Zhao F, He J, Li X, Bai Y, Ying Y and Ping J 2020 Smart plant-wearable biosensor for *in situ* pesticide analysis *Biosens. Bioelectron.* **170** 112636

[53] Wang S *et al* 2020 Advanced triboelectric nanogenerator with multi-mode energy harvesting and anti-impact properties for smart glove and wearable e-textile *Nano Energy* **78** 105291

[54] Shuai L, Guo Z H, Zhang P, Wan J, Pu X and Wang Z L 2020 Stretchable, self-healing, conductive hydrogel fibers for strain sensing and triboelectric energy-harvesting smart textiles *Nano Energy* **78** 105389

[55] Veeralingam S, Khandelwal S, Sha R and Badhulika S 2020 Direct growth of FeS_2 on paper: a flexible, multifunctional platform for ultra-low cost, low power memristor and wearable non-contact breath sensor for activity detection *Mater. Sci. Semicond. Process.* **108** 104910

[56] Reddy K C S, Selamneni V, Rao M S, Meza-Arroyo J, Sahatiya P and Ramirez-Bon R 2021 All solution processed flexible p-NiO/n-CdS rectifying junction: applications towards broadband photodetector and human breath monitoring *Appl. Surf. Sci.* **568** 150944

[57] Li S, Li R, González O G, Chen T and Xiao X 2021 Highly sensitive and flexible piezoresistive sensor based on c-MWCNTs decorated TPU electrospun fibrous network for human motion detection *Compos. Sci. Technol.* **203** 108617

[58] Li T *et al* 2020 A flexible strain sensor based on CNTs/PDMS microspheres for human motion detection *Sens. Actuators* A **306** 111959

[59] Wang J *et al* 2021 A stretchable self-powered triboelectric tactile sensor with EGaIn alloy electrode for ultra-low-pressure detection *Nano Energy* **89** 106320

[60] Wu G *et al* 2021 High performance stretchable fibrous supercapacitors and flexible strain sensors based on CNTs/MXene-TPU hybrid fibers *Electrochim. Acta* **395** 139141

[61] Almukhlifi H A, Khasim S and Pasha A 2021 Fabrication and testing of low-cost and flexible smart sensors based on conductive PEDOT-PSS nanocomposite films for the detection of liquefied petroleum gas (LPG) at room temperature *Mater. Chem. Phys.* **263** 124414

[62] Xue H and Hu J 2021 A liquid power-ultrasound based green fabrication process for flexible strain sensors at room temperature and normal pressure *Sens. Actuators* A **329** 112822

[63] Hossain G, Hossain I Z and Grabher G 2020 Piezoresistive smart-textile sensor for inventory management record *Sens. Actuators* A **315** 112300

[64] Shen Y, Yang F, Lu W, Chen W, Huang S and Li N 2021 A highly stretchable and breathable polyurethane fibrous membrane sensor for human motion monitoring and voice signal recognition *Sens. Actuators* A **331** 112974

[65] Tian G *et al* 2021 Coating of multi-wall carbon nanotubes (MWCNTs) on three-dimensional, bicomponent nonwovens as wearable and high-performance piezoresistive sensors *Chem. Eng. J.* **425** 130682

[66] Khalifa M, Wuzella G, Lammer H and Mahendran A R 2020 Smart paper from graphene coated cellulose for high-performance humidity and piezoresistive force sensor *Synth. Met.* **266** 116420

IOP Publishing

3D Printed Smart Sensors and Energy Harvesting Devices
Concepts, fabrication and applications
Sanket Goel and Sohan Dudala

Chapter 11

Applications: energy harvesting and storage

Valentin Mateev

This chapter considers different technologies for energy harvesting and storage applications, produced by 3D printing technologies and used for small sensing device powering. A review of energy harvesting methods, from accessible environmental sources such as electromagnetic fields and waves, mechanical vibration, and light, temperature, and pressure differences, applied via 3D-printed devices and materials, is presented. A comparison of the limitations of the current state-of-the-art is provided, and future technological trends and boundaries are highlighted. Next, possible new methods for energy storage in microdevices, which are realizable by innovative additive technologies, are presented. Energy storage technologies in small Internet of Things sensor devices cover not only batteries but also many other methods, such as deformation, inertial, gravitational, pressure, etc, accumulation technologies, the application of which can be supported by new additive technologies. The energy and power consumption of small sensor devices are considered according to their frequency of use, data transmitting interfaces, and operational life expectancy. Some indications of the direct use of batteries for energy harvesting are discussed, and the possible future for 3D-printed galvanic cells beyond the limitations of Faraday's electrochemical equivalents at the microscopic scale is discussed.

11.1 Introduction

3D printing technologies are becoming a real game-changer in applied designs and have gained accessibility in local communities at a global scale. The wide distribution of 3D printing machines and materials is currently creating the basis of a new manufacturing revolution that is just starting to emerge in developing countries where they have no existing competition from classical mass production methods. The renewable energy transition towards sustainable local societies is creating new requirements for power supply through innovations in energy acquisition and storage. Preferably, new sustainable energy conversion and storage

doi:10.1088/978-0-7503-5351-9ch11
11-1

equipment should be manufactured by additive technologies close to the end-user's site and using local materials, implementing optimal designs for the specific applications required. Small electronic devices, autonomous sensor nodes, and smart instruments will be the obvious target objects for new 3D-printed energy harvesting and storage designs. The recent development of high-tech 3D printing technology and material science must be translated into the language of the local users of 3D printing, empowering them to use innovations at the edge of their technological boundaries as an engine for a sustainable transition.

The content of this chapter is structured by starting with an introduction to the problem, followed by the basic theory of energy harvesting and design concepts. The next section contains an estimation of the energy consumption of Internet of Things (IoT) devices as the main objective of energy harvesting, followed by an overview of some recent 3D printing technologies and materials for energy harvesting systems. The chapter provides a review of the energy storage capabilities of 3D-printed designs, and the final section is the conclusion.

11.2 Basic principles of energy harvesting

With the fast development of the concept of autonomous distributed electronic devices, energy harvesting is becoming more important. IoT distributed devices require electrical energy for sensing, data processing, and data transmission in interconnected autonomous networks. With the fast development of the technology of IoT devices, the required power consumption is reduced and many new energy harvesting devices and principles are becoming accessible for such applications. From another perspective, low-cost energy harvesting solutions with a targeted use could have a significant impact as a practical onsite technological boost for autonomous and wearable devices. Three-dimensional printing technologies are becoming a key tool for such innovations.

Free energy sources such as temperature differences, vibrations, different movements, light, sound, pressure changes, and humidity are naturally available as direct energy for small sensors. New harvesting designs optimize the receiving process and increase the efficiency of energy absorption. A quantitative estimation was made of each energy harvesting approach supplied by the primary energy source. Recommendations for designing and prototyping energy harvesting devices are highlighted here. Promising device designs are reviewed in the next subsection.

The basic general theory of energy harvesting can be summarized as five main principles:
1. There is a free primary power source outside the energy harvesting device.
2. If the power flowing from the source is diffuse, it must be converted into a directed power flow.
3. Directed power flow from the source must interact with the harvesting device's receiver.
4. To absorb energy from the source, the internal resistance of the receiver must be smaller than that of the primary source.
5. Some of absorbed energy can be extracted and used for useful work.

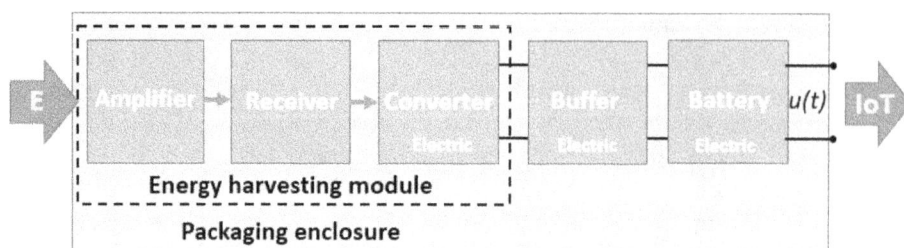

Figure 11.1. Block scheme of an energy harvesting module.

To implement these five principles, energy harvesting devices have a common functional structure, as shown in figure 11.1.

A block scheme of an energy harvesting device is presented in figure 11.1. It represents a functional scheme which may include an input amplifier, an active receiver element, and an electric converter. After the electric converter block, the power flows into an outer electric buffer and/or a long-term battery for electric energy storage. The key element in the energy harvesting system is the active receiver element, which interacts with and receives the external physical input (electromagnetic waves, sound waves, vibrations, temperature gradient, pressure gradient, etc) and converts it to usable electric power. Different kinds of input amplifiers may be used to focus the outer physical field into the active receiver element's domain. Thus the harvested power level is increased. Input amplifiers may be focusing systems used as waveguides for wave harvesters, or collimators, mirrors, and lenses for optical harvesters. The active receiver element couples with the accessible outer power source and directs the flow to the electric converter for generating electric voltage.

Each of these elements of the energy harvesting module (figure 11.1) can be produced by 3D printing technology.

The primary sources and the expected range of energy harvested are presented in table 11.1. A volume of 1 cm^3 was chosen as a comparative size domain. We assumed that this volume was available to integrate the standalone harvesting device and was further used as a power supply for the IoT sensor controller. The harvested power depends on many environmental conditions, but the possible maximum power harvest of the current technology is shown in table 11.1.

The next section indicates some trends of development, and the current and future capabilities of 3D printing technology for making energy conversion devices.

11.3 Energy consumption of IoT sensing nodes

Most of the desired applications of energy harvesters are for IoT sensing nodes. They must be autonomous for long enough to eliminate the power supply network and to allow flexible operation. With advances in the optimization of power usage, more and more electronic portable devices, such as tablets, smart gadgets, etc, fall into the categories of possible energy harvesting power supply or supportive power

Table 11.1. Frequency spectra and specific powers.

Energy source/EH power flow	Design	Frequency	Power harvest expected	Applicability to 3D printing
Electromagnetic RF waves	Antenna, heat absorption	kHz–GHz	0.01–10 mW cm^{-2}	Very high
Solar light or diffuse light	Solar cell, solid dilatory, fluid exchange, heat absorption	>THz	0.1 W cm^{-2}	Partial
Vibration and sound	Resonant, electric, magnetic	kHz	1 μW cm^{-2}	Partial
Kinetic, human or other movement	Inertial, electric, magnetic	Hz	0.5 mW cm^{-3}	High
Temperature difference	Direct thermoelectric	NA	0.1 mW cm^{-3} Temp dependent	Partial
Pressure (air or contact)	Fluid dilatory, deformation, tactile	NA	20 μW cm^{-2}	Yes
Humidity	Galvanic, absorption	NA	1 nW cm^{-2}	Yes

supply. Light-emitting diodes are a good example of the minimization of energy consumption and the advanced efficiency of energy conversion. Imagine tiny self-powered Christmas lights shining in the night to visualize the possibilities of distributed autonomous sensor networks. The main sources of consumption of an autonomous IoT sensing node are presented below.

The estimated peak power consumption of a sensor node includes the power supply for the controller unit (up to 330 mW; 100 mA at 3.3 V), consumption by the sensor (200 mW), and the transmitter's power (LoRa case, 150 mW). The transmitter's power usage is limited only to the transmission interval, which is one every minute, so the energy consumption is very low for LoRa. The majority of the energy is consumed by the controller, which is a subject for optimization. An IoT sensor node device is powered by a stack of lithium batteries to cover the power consumption of the controller's MCU, the sensor, and the transmitter. The battery life of the sensor node device is highly dependent on the refreshing time of acquisition and the signal spreading factor used. Higher spreading factors result in longer active times for the radio transceivers and a shorter battery life. It must be pointed that the power consumption of the sensor used for measurement and the sending power of the transmitter are limited to short time periods, so the total amount of energy is low.

The consumption of portable IoT devices supported by wireless communication with gateway devices supporting the integrity of communication is very important. The average power consumption of an idle equivalent IoT device is shown in table 11.2. It must be noted that the peak power could not be covered by any energy harvesting receiver; therefore, energy buffering accumulators must be used [1].

The transmitter's sending power is limited to short time sessions, so the total amount of energy is low. The time on air (ToA) depends on the distance of transmission and on the number of measuring nodes. When the distance of transmission increases, the data rate has to be decreased accordingly because of the receiver's signal-to-noise ratio. Increasing the number of devices also results in a larger scanning cycle and many more back-transmissions per device.

An example of large power usage by a sensor is provided by the catalytic gas sensor MQ-135, built as a resistive Wheatstone bridge circuit. A reference bead is adjusted to maintain a state of electric balance in the bridge circuit in clean air. The measured gas concentration will affect the detector's active bead resistance, which will rise, causing an imbalance in the bridge circuit. The MQ-135 sensor is supplied with a voltage of 5 V at 50–200 mA current for consumption. Only the sensor's heating requires a continuous power supply, typically around 200–800 mW. This huge amount of power is not required by the new generation of passive absorption sensors. Therefore, most power losses are caused by the heater's resistance ($R_h = 33$ Ω); the balanced bridge's equivalent resistance (R_s) is 30 kΩ. The consumption of the sensor circuit, without the consumption of the heater, is 1 mW. Variation in the sensor's resistance affects the signal of the drop in the output voltage, which is proportional to the gas concentration. Because of the slow changes in the gas concentration, most IoT environmental monitoring systems activate the sensor circuitry only at the desired moment of data acquisition. This is possible in real-time measurements where the no-current pause is larger than the sensor activation period. Here, data acquisition takes place in 20–30 s intervals, which is less than the sensor's heating constant. This significantly limits the sensor's operational life [1].

Table 11.2. Average power consumption of equivalent IoT devices.

Mode →	Power on standby	Continuous power	Peak power
Wireless interface ↓	μW	mW (avg. per 1 s)	mW
WiFi	150	15	1000
Bluetooth	30	5	120
LoRaLAN	95	10	150
ZigBee	45	7	160

It must be noted that the peak power of an IoT sensing node cannot be covered by any energy harvesting receiver; therefore energy buffering accumulators must be used for voltage stabilization during the time of usage of the IoT sensing node.

11.4 3D printing technologies and materials for energy harvesting and storage

This section mainly considers fused deposition modeling, also known as fused filament fabrication (FDM/FFF), and stereolithography (SLA) 3D printing technologies, as these are the most widespread and accessible for large groups of users [2, 3].

Fused filament fabrication 3D printing technology was originally developed for polymer thermoplastics such as ABS and, later, the famous PLA material. These polymers are excellent electrical insulators, with very well studied material properties. Different additives and fillers change the solid polymer's base properties to improve certain mechanical, thermal, magnetic, or electric properties. These so-called composites exhibit a wide variety of material properties that can be applied for solving many engineering design problems. The volumetrically controlled electric conductivity of 3D-printed materials is one of most desired properties in electrical engineering applications. Such composites are directed for use in low-voltage circuits, numeric keypads that require low conductivity, deformation sensors, robotics, and flexible electronics [3–5].

SLA is faster and much more energy-efficient than fused deposition modeling. The single material reservoir in SLA usage is a huge barrier in forming composite materials with varying added layers or laminations. However, volumetric dispersed compositions are still possible to be formed [6]. SLA printing of parts has better production speed, resolution, energy consumption and accuracy, compared with thermoplastic fused deposition modeling. Material manufacturers have created innovative SLA resin formulations with a wide range of optical, mechanical, and thermal properties to match those of standard, engineering, and industrial polymer compositions.

FDM/FFF 3D printing materials need to exceed the material melting point in the extruder. This is required only for the carrier material in the composites; for example, metal or non-organic fillers are not melted but are only mixed in the main compound. SLA is more temperature-friendly, reaching lower temperatures during the UV curing process in the trim surface layers.

An overview of FDM/FFF and SLA 3D printing materials' properties is presented below.

Mechanical properties. The mechanical strength of PLA is typical of polymer plastic compositions, which is highly acceptable for the devices applied in modern life. A typical value for PLA's tensile strength is 40 MPa and its flexural strength is 80 MPa. With additives and filling patterns, these material properties can be modified. ABS has the same range of mechanical strength.

Electric insulation properties. PLA is an excellent dielectric polymer plastic. Its relative permittivity is in the range of $\varepsilon_r = 2$–7 and the loss tangent is >1e−4, with a break electric field intensity of 40 kV mm^{-1} for a 100% material fill ratio. Dried SLA resins also have approximately these values.

The electric conductivity of polymers that are subject to FDM/FFF and SLA 3D printing is extremely low; these are dielectric materials. Some conductive additives are changing this, but electrical conductivity remains low for most power applications. Therefore, integration with metal alloys with a low melting point would be a useful direction of research for FDM/FFF 3D printing of wires, traces, and coils. This way, the electrical conductivity of solid metal could be implemented in polymer assemblies, which are insulators [4].

Magnetic permeability and coercivity. FDM/FFF technology has limited use in printing magnetic material. Many research groups are working on developing different compositions, based mainly on metal powders [4], for the production of soft and hard magnetic filaments [5]. These new composite materials are currently far from the average expectations of solid metal magnetic materials, both in terms of magnetic properties and cost, but they can be very useful in many applications for rapid prototyping in electronics and power conversion, e.g. for different inductors, HF antennas, chokes, sensors, transformers, new elastomer electromagnetic actuators, and flexible electronics.

The *optical properties* are not so good and are limited by the layers' thickness and surface roughness after 3D printing. This is not satisfactory for optical mirrors and lenses. For waveguides, the roughness of the 3D printing z-layers limits the wavelength to above 100 μm, which is the microwave range in air. PLA has higher dielectric permittivity. Microwaves are outside the visible light range (400 and 700 nm), but overlap enough in the far-infrared spectrum (15 μm to 1 mm) (table 11.3).

Most of an FDM/FFF 3D printer's extruders operate at temperatures of up to 450 °C without changes in the design and controller. Soft metals (e.g. lead (Pb)) and metal alloys (containing lead (Pb), bismuth (Bi), tin (Sn), or gallium (Ga)) can be used in the printing process as metal filaments. Pb has a melting point of 327.5 °C, and the melting point is 271.5 °C for Bi and 30 °C for Ga. The main material

Table 11.3. Material properties.

Type	Material properties	Value
Mechanical	Tensile strength	60 MPa
	Young's modulus	4 GPa
	Impact brittle strength	90 J m^{-1}
Electric	Permittivity ε_r	1–10
	Conductivity* σ	0.1–5 S m^{-1}
Magnetic**	Permeability μ_r	1–10 [5]
	Coercivity H_c	>50 kA m^{-1}
Thermal	Thermal conductivity λ	0.13 W m^{-1} K^{-1}
	Thermal capacity c	1.6 kJ kg^{-1} K^{-1}
Optical***	Absorption (visible light) μ_s	0.1
	Index of refraction	1.4–1.5

*For conductive PLA polymers.
**For magnetic PLA compositions.
***For transparent PLA.

Table 11.4. Material properties.

	Pb	Bi	Ga	Sn	PLA conductive
Melting temperature	327.46 °C	271.5 °C	30 °C	231.93 °C	230 °C
Density	11.34 g cm^{-3}	9.78 g cm^{-3}	5.91 g cm^{-3}	6.99 g cm^{-3}	1.24 g cm^{-3}
Bulk electric resistivity	208 nΩ m (at 20 °C)	1.29 $\mu\Omega$ m (at 20 °C)	270 nΩ m (at 20 °C)	115 nΩ m (at 0 °C)	200 Ω m
Thermal conductivity	35.3 W m^{-1} K^{-1}	7.97 W m^{-1} K^{-1}	40.6 W m^{-1} K^{-1}	66.8 W m^{-1} K^{-1}	0.15 W m^{-1} K^{-1}
Heat capacity	26.650 J mol^{-1} K^{-1}	25.52 J mol^{-1} K^{-1}	25.86 J mol^{-1} K^{-1}	112 J mol^{-1} K^{-1}	99 J mol^{-1} K^{-1}
Shear deformation modulus	5.6 GPa	12 GPa	<10 GPa	18 GPa	3.5 GPa

properties are listed in table 11.4. These metals have excellent electrical conductivity compared with copper (Cu) (17.7 nΩ m), significantly better than the best electrically conductive PLA composites.

Single-metal FDM/FFF 3D printing is applicable for standalone elements, which can be later used for assemblies. To assemble the elements of electrical device during printing, two-component printing is needed. Two components will provide material properties with an operational gradient, such as a conductive wire in an insulated enclosure or cover. For that purpose, for thermal stability during FDM/FFF printing, the second material must have a lower temperature than the base printing material (PLA, ABS). This way, for example, 3D-printed metal wires with complex shapes can be integrated into the enclosed polymer design. Such low-temperature metal alloys exist (e.g. Wood's metal, with a melting point of 70 °C) and are freely applicable for FDM/FFF volumetric formation with appropriate processing parameters. Wood's metal is a eutectic alloy of 50% bismuth, 26.7% lead, 13.3% tin, and 10% cadmium. It has an attractive (from an engineering point of view) yield strength of 26.2 MPa [7]. Similar alternatives are Rose's metal alloy and Field's metal alloy. The electric conductivity of metal alloys is not comparable with polymer composites; it is million times greater than the best of them, but ten times lower than that of copper alloys. Low-temperature alloys for FDM/FFF technology need additional research to reveal the optimal control parameters during 3D printing and to achieve technological experience.

11.5 3D three-dimensional printable energy harvesting designs

This section mainly considers FDM/FFF and SLA 3D printing technologies as the most widespread and accessible for large groups of users. Energy harvesting devices capture and convert small amounts of energy from the outer environment. This energy comes from the dissipation of larger energy systems, moving objects, or specific environmental events. Harvested energy is considered as free energy that is enough for powering very small autonomous devices. The sources of energy are not reliable, and the amount harvested can vary over time. Here, we describe the principles of the electromagnetic, vibration, kinetic, temperature, and pressure difference technologies that can be used by 3D printing implementations. The limitations of the technology of each method are indicated.

Electromagnetic energy harvesting devices can be separated into low-frequency and high-frequency types [8]. Depending on the field strength, coils and capacitor plates are used in low-frequency devices to induce electric voltage as a power supply. The Earth or another magnetic field (40 μT) can induce in the required amount of power in a moving coil. This field can be focused by ferromagnetic cores and concentrators to enlarge its productivity. In a similar way, dielectric materials, such as clothes, upholstery, resins and polymers, can store an electric charge that can be used as a for power supply. The current low-cost 3D printing technologies (FDM/FFF and SLA) are not fully optimized for electrically conductive materials with high conductivity, so they are not able to produce good low-resistivity coils (in the case of magnetism) or contacting electrodes (in the case of electricity). However, some alloys with a low melting temperature can be used by FDM/FFF technology for that purpose, creating low-resistance coils and electrodes [9]. A special case of magnetic harvesters includes those that use movable magnets for voltage induction; these are considered in the section on kinetic harvesters. Electric tribological, bendable, and stretchable harvesters are also described in the section on kinetic harvesters.

High-frequency electromagnetic harvesting relies on the absorption of RF waves by planar antennas. Scattered waves from communication networks in cities could provide a few milliwatts to an antenna with a 1 cm^2 surface. Such 3D-printed antennas are easily implementable with FDM/FFF conductive materials over a dielectric substrate [8–11]. Three-dimensional printing has a good ability to produce RF harvesters with directional selectivity, which is not easily achieved with planar foil antennas. This way, volumetric microwave lenses can be integrated directly in the antenna's design. These microwave lenses are based on designs with gradient properties or multilayers with adaptive density [12, 13]

Solar light or diffuse ambient light can be treated as electromagnetic energy with a very high frequency, above the THz bandwidth. The expected energy flow is not larger than 50 mW per 1 cm^2. The Sun is the most powerful concentrated energy source for harvesting; unfortunately, it is locally usable only during the day [14–16]. Solar energy harvesting works not only under direct sunlight but also under the power of LEDs or florescent light sources for in-house applications. Photovoltaic (PV) semiconductor cells are not suitable for direct 3D printing technology, but some light-absorbing materials could be applied and shaped by 3D printing. Perovskite-enriched substrates [17] or composites in a substrate with low electrical conductivity could be used to produce solar cells. Moreover, some semi-transparent filaments and resins could be used for fabricating light-concentrating lenses for increased light intensity. Unfortunately, the materials' optical transparency is not significant, and lenses create very high directional selectivity that reduces the time of usage of the solar cell during the day. Three-dimensional printing technologies are usable for solar mirror concentrators; in that case, metallization coatings are suitable for creating convex reflective surfaces. For FDM/FFF-printed mirror substrates, a pre-polishing procedure is needed because of the high surface roughness of directly printed elements [18, 19].

Vibration and sound energy harvesting. In mechanical vibration energy harvesters, the active element is typically a free pendulum or elastic plate (for ultrasound frequencies) and movable mass particles (for high-vibration frequencies). Three-dimensional printing methods and materials are really useful for making receiver elements such as pendulums or plates. The converter can be piezoelectric or magnetostrictive. The conversion efficiency of piezoelectric modules is high, with a low conversion loss. Piezoelectric modules cannot be considered for manufacturing by additive technologies. Iron microparticles and ferrofluids are considered to be the active elements of the energy harvester, providing continuous movement in the electricity harvesting coils, thus inducing a usable voltage.

Kinetic harvesters use human or other movements, which are intensified by resonant systems. These systems increase the frequency, making them relatively stable and limiting the amplitude of deviation. The moving parts (movers) of kinetic harvesters enlarge the relaxation time, increasing the exposition time for harvesting energy. Kinetic energy harvesters are typically built from an inertial mass element and an elastic spring with a constant stiffness. Three-dimensional printed elements are very suitable for printing such movers. Movers can have planar–linear or circular oscillation. Circular oscillations are more compact, while planar–linear oscillations have better selectivity for detecting frequencies. The size of the planar–linear mover plates determine the resonant frequency of the harvester. The mover's tip is typically a permanent magnet, so vibrations are converted directly to induced electric voltage in a nearby coil. Opportunities for 3D printing high-conductivity electric coils were considered in the previous section.

Temperature differences are among the most attractive energy sources. Almost everywhere, we have static and dynamic temperature differences from different origins. These temperature gradients produce power flows that are directly captured and converted to electricity by thermoelectric Peltier generators (TEG) [20, 21]. Infrared emissions can be treated in a similar way to the light absorbed by PV cells or direct THz diode arrays. In both cases, 3D printing is only applicable for concentrating the infrared light produced by refraction because 3D-printed polymers and metal materials are not transparent to IR light. For TEGs, 3D printing is applicable for manufacturing pre-concentrators for snow, water, or other temperature-transferring materials with a high thermal capacity [20, 21].

Contact pressure energy harvesting relies on relatively static pressures. It uses a force striking a perceptive element. The frequency of the pressure of such strikes is low (1 Hz). The perceptive element can be a permanent magnet in a coil or a piezoelectric crystal plate. Typically, these harvesters include self-powered keyboards and active pavements. These harvesters, especially their perceptive elements, show huge potential for 3D printing and wide design diversity. The harvest is very low for moving coils. Other contact and deformation designs include bendable and stretchable membrane harvesters [26].

Friction (tribological) electricity harvesters. Triboelectric charging of dielectrics by surface friction is a simple and reliable way of charging a device. Relative rotational and translational movements in many devices produce a voltage potential that can be adopted for a low power supply. They are especially suitable for 3D printable polymer dielectric materials with very low leakage currents that keep their electric charge for a long time. In some cases, protection against overvoltage must be foreseen. These applications are most desirable in wearable electronics, clothes that integrate sensors, and cars moving over dry roads [22, 23].

Air pressure harvesters. Daily changes in atmospheric pressure have been used since the times of mercury thermometers. Some early mechanical clocks were powered this way from the 1700s. Nowadays, gallium is used as an excellent replacement for mercury and falls within the temperature range of FDM/FFF 3D printing. Gallium can fill a polymer microchannel during polymer printing and realize a complete 3D-printed assembly with good sealing from the outer environment, except for the air pressure perception membrane [24–26].

Mechanical wave vibration harvesters are similar to contact pressure harvesters but have a higher frequency range (100 Hz–1 MHz). To absorb high-frequency vibrations beyond 5 kHz, light particles are used for conversion, because of their small mass [27–30]; fluids have also shown encouraging results. Three-dimensional printing by FDM/FFF and SLA is extremely suitable for making such transducers.

Humidity harvesters. Harvesting involves not only the absorption of a physical impulse; in some cases, a material resource is absorbed and used. These can be organic vapors or organic liquids (e.g. ethanol, ammonia, or alcohol) that are free-floating and can be captured from the outer environment. Moreover, water coming from rain, the soil, or humid air is usable as an activator of electrochemical reactions. Rusty galvanic cells (called rusty batteries) are very suitable for environments with high humidity. Rusty batteries are perfect for direct manufacturing from conductive iron-filled PLA composites [31].

Electrochemical harvesters. As a harvester of a material resource, glycose cells in implantable sensors for *in vivo* applications (called sugar batteries or sugar cells) can provide small amounts of electric power resulting from oxidizing polarization (glycolysis) in a microchannel between a pair of electrodes [32–34]. The caloric oxidation energy value of pure concentrated glucose is 15.7 kJ g^{-1}, but only a very small amount of it can be converted into electricity. The input source also comes in very low concentrations. Glucose-containing filaments are suitable for 3D FDM/FFF printing at low processing temperatures. Metal electrodes and polymer compositions are applicable for such chemical harvesting designs [35].

Chemically active 3D-printed materials. These applications subject to chemical degradation, and irreversible or reversible changes during their operational lifetime. This is important for galvanic cells, as considered in section 11.6.

To absorb the environmental energy, the internal resistance of the receiver must be smaller than that of the primary source. Energy comes in short pulses, with a variable duration, and the power consumption of IoT sensor nodes does not have an even distribution over time, so electrical energy must be accumulated before the usage of harvested energy. In most cases, a battery or capacitor is required as an energy buffer in energy harvesting systems to avoid interruptions in the absorbing process. A battery is required not only as an energy buffer in energy harvesting, but it is also applicable as a primary energy harvesting element. The energy of the outer environment, such as a thermal gradient, pressure, etc, could be used for intensifying the internal reaction of the battery cells. Such applications of temperature, vibration, and electrochemical cell stimulation could be extremely useful for specially design battery elements that are open to the outer environment.

11.6 Energy storage by 3D-printed designs

Energy storage in small IoT sensor devices does not apply only electrochemical batteries, but many other methods (deformation, inertial, gravitational, pressure, etc) that can be freely applied or supported by new additive technologies. This section considers mainly FFF/FDM and SLA printing technologies as the most widespread and accessible for large groups of users.

A battery is required as an energy buffer in energy harvesting. Typically, it is assumed that 3D printing is not suitable for the formation of thin foils, but it is excellent for making 3D pouring foams and volumetrically mixed cord threads. The potential of 3D printable electrodes for electrochemical cells is enormous and cannot be fully covered in this chapter. However, some tendencies have been noted as innovative and prospective technological directions with promising benefits.

A block scheme of the general structure of a battery cell is presented in figure 11.2. It consists of positive (anode) and negative (cathode) electrodes, the electrolyte, and the separator. Battery cells also include connection electrodes and enclosures, depending on the particular design [36–40].

FFF and STL 3D printing are not the optimal methods for producing thin foil materials, which are mainly used for making battery cells. New volumetric electro-chemical cells provide a high specific reaction surface with a small volume. Printing volumetric repeating patterns with microscale distances between the electrodes can increase the cell's capacity. Even the famous spaghetti printing error of FFF printing

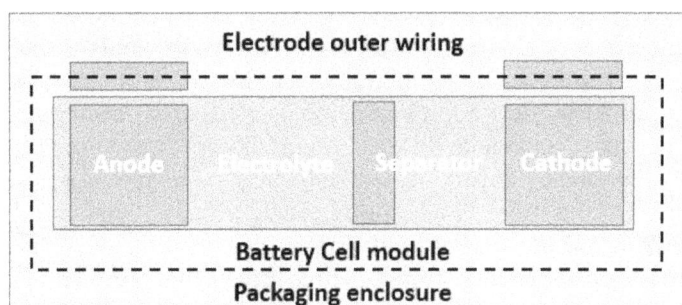

Figure 11.2. Main elements of a battery cell, suitable for 3D printing.

without contact with the base layers is considered to be an opportunity for the creation of volumetric cloud electrodes.

Low-temperature metals have also been considered for making electrodes. The traditional Pb electrodes, used for lead–acid batteries, are printable by FFF, and are providing huge variations in the filling patterns for enhanced electrolyte contact. Similar approaches have been applied successfully for Ga, Na, and Zn battery electrodes [36–39].

The use of diverse composite materials for FFF 3D printing provides some insights for electrochemistry beyond Faraday's electrolysis mass ratios. The common use of Faraday's electrolysis mass ratios is in calculations of electrolysis or electrode degradation. It defines the experimentally determined charge (or electric current) and the chemical amount of a substance (in moles) that has been electrolyzed [32–34]. Not all the electric current is used for dissolution of the anode. The current's efficiency is defined as the ratio of the observed amount of metal dissolved to the theoretical amount predicted by Faraday's law. Faraday's mass ratios of electrolysis are experimentally determined for metal ions. Three-dimensional printable filament compositions with metal additives or transition metal additives make separable secondary charged electrical particles that are beyond Faraday's electrochemical equivalent mass ratios for pure metal ions in a near electric field. These molecular particle complexes or secondary particle constellations can carry larger charges, consuming or storing energy, while transporting them in an external electric field [39–41].

Batteries' electrochemistry could be subjected to stimulation, especially polymer materials [41–46], thus providing a harvesting effect during energy storage. The energy from the external environment, such as thermal gradients, pressure [39], etc, could be used for intensifying the internal reactions of a battery cell, increasing its charge without outer power supply. Such applications of temperature and vibration for electrochemical stimulation of a cell could be extremely useful with specially designed battery elements that are open to the external environment as air-breathing batteries [41, 49].

The state-of-the-art of FDM/FFF and SLA printing does not provide stable technical solutions for proton-exchange membrane manufacturing for fuel cells. There are some indications that microfluidic fuel cells can be produced by SLA, so this is a positive indication for the reliable printing of fuel cell membranes in the future [47–49].

Energy storage in small IoT sensor devices does not apply only to electrochemical batteries, but many other methods (deformation, inertia, and others) can be freely applied or supported by new additive technologies. Inertial accumulators for IoT sensor nodes produced as kinetic flywheel accumulators have many difficulties at a small scale, but bearings and friction problems may be solved in a different way in the near future. On the other hand, spring elastic accumulators could be adopted from watchmaking, where tension and torsion deformation of a linear coil or a circular spring is used for energy storage. Three-dimensional printed metal filaments are directly applicable for spring accumulators. The only problem is the small energy density of the springs made from these low-temperature metals [48].

11.7 Conclusion

The advances in 3D printing are able to overcome the main barriers to optimizing energy consumption and material efficiency (resource efficiency) in industrial settings at the end-users' point of manufacture. Small electronic devices, autonomous sensor nodes, and smart instruments are the obvious target objects for new 3D-printed energy harvesting and storage designs. Three-dimensional printing by FDM/FFF and SLA is good for elements of energy perception and supporting and supplementary elements rather than for printing electrical conversion elements.

Some trends for future development can be highlighted as follows:
- The advances in energy harvesting achieved by 3D printing have mainly focused on new materials rather than fundamentally new device designs.
- Low-temperature metal alloys could be used for integrating wires and electrodes in 3D polymer printing. Some of these metal alloys have good elasticity properties and are usable for oscillating springs and plates.
- The rise of semiconductor additives in 3D polymer printing will open many doors for low-power energy conversion.
- Biocompatible materials will allow the integration of implants for interactive cyber–physical systems.

The best applications of 3D printing for energy conversion and storage at small scales for IoT sensors still remain to be found.

References

[1] Mateev V and Marinova I 2021 ANN for electromagnetic design optimization *12th Nat. Conf. with Int. Participation ELECTRONICA* pp 1–5
[2] Migliorini L *et al* 2022 Nanomaterials and printing techniques for 2D and 3D soft electronics *Nano Furtures* **6** 032001
[3] Aliheidari N, Hohimer C and Ameli A 2017 3D-printed conductive nanocomposites for liquid sensing applications *Conf. on Smart Materials, Adaptive Structures and Intelligent Systems* vol 1 (New York: ASME)
[4] Ralchev M, Mateev V and Marinova I 2021 3D Printed Electrically Conductive Composites by FFF/FDM Technology *13th Electrical Engineering Faculty Conf. BulEF*
[5] Ralchev M, Mateev V and Marinova I 2021 Magnetic properties of FFF/FDM 3D printed magnetic material *17th Conf. on Electrical Machines, Drives and Power Systems (ELMA) (Sofia)*
[6] Ralchev M, Mateev V and Marinova I 2022 3D printed polymer composite magnetic material by stereolithography technology *22nd Int. Symp. on Electrical Apparatus and Technologies (SIELA) (Bourgas)*
[7] White G K 1987 *Experimental Techniques in Low-temperature Physics* 3rd edn (Oxford: Oxford University Press)
[8] Wang T, Huang K, Liu W, Hou J and Zhang Z 2022 A hybrid solar-RF energy harvesting system based on tree-shaped antenna array *Int. J. RF Microw. Comput. Aided Eng* **32** 10
[9] Kahar K *et al* 2022 MEMS-based energy scavengers: journey and future *Microsyst. Technol.* **28** 1971–93

[10] Wang Z, Huber C, Hu J, He J, Suess D and Wang S 2019 An electrodynamic energy harvester with a 3D printed magnet and optimized topology *Appl. Phys. Lett.* **114** 013902

[11] Takemura N, Kohara Y, Ichikawa S and Kondo H 2017 A study of directional antenna for recycled energy improvement in electromagnetic wave energy harvesting *IEEE Int. Symp. on Antennas and Propagation and USNC/URSI Nat. Radio Science Meeting*

[12] Bedair S *et al* 2021 Piezoelectric and ferroelectric devices for energy efficiency and power *IEEE Int. Symp. on Applications of Ferroelectrics (ISAF) (Sydney)*

[13] Zhang C, Zhou Y J, Xiao Q X, Yang L, Pan T Y and Ma H F 2016 High-efficiency electromagnetic wave conversion metasurfaces for wireless energy harvesting *Progress in Electromagnetic Research Symp. (PIERS) (Shanghai)*

[14] Yang Z, Jia S, Niu Y, Lv X, Fu H, Zhang Y, Liu D, Wang B and Li Q 2021 Bean-pod-inspired 3D-printed phase change microlattices for solar-thermal energy harvesting and storage *Small* **17** 2101093

[15] Tol S, Degertekin F L and Erturk A 2017 3D-printed lens for structure-borne wave focusing and energy harvesting *Active and Passive Smart Structures and Integrated Systems (Portland, OR)* (SPIE) 10164

[16] Nauroze S A *et al* 2016 Additive manufacturing technologies for near-and far-field energy harvesting applications *IEEE Radio and Wireless Symp. (RWS) (Austen TX)*

[17] Christians J A, Marshall A R, Zhao Q, Ndione P, Sanehira E M and Luther J M 2018 Perovskite quantum dots. A new absorber for perovskite–perovskite tandem solar cells *IEEE 7th World Conf. on Photovoltaic Energy Conversion (WCPEC) (Waikoloa, HI)*

[18] Wang Z, Zhan Z, Chen L, Duan G, Cheng P, Kong H, Chen Y and Duan H 2022 3D-printed bionic solar evaporator *Sol. RRL* **6** 2101063

[19] Melgarejo V, García L, Reifenberger J and Newell B 2018 Manufacture of lenses and diffraction gratings using DLP as an additive manufacturing technology *Proc. of the ASME Conf. on Smart Materials, Adaptive Structures and Intelligent Systems* 2

[20] Peng J *et al* 2019 3D extruded composite thermoelectric threads for flexible energy harvesting *Nat. Commun.* **10** 5590

[21] Noh Y S *et al* 2021 A reconfigurable DC-DC converter for maximum TEG energy harvesting in a battery-powered wireless sensor node *IEEE Int. Solid- State Circuits Conf. (ISSCC) (San Francisco, CA)*

[22] Adhikari P R, Islam M N, Jiang Y, Reid R C and Mahbub I 2022 Reverse electrowetting-on-dielectric energy harvesting using 3-D printed flexible electrodes for self-Powered wearable sensors *IEEE Sens. Lett.* **6** 6001704

[23] Gowthaman S *et al* 2018 A review on energy harvesting using 3D printed fabrics for wearable electronics *J. Inst. Eng. India Ser.* C **99** 435–47

[24] Yuan M, Cao Z, Luo J and Pang Z 2018 Helix structure for low frequency acoustic energy harvesting *Rev. Sci. Instrum.* **89** 055002

[25] Kawa B and Walczak R 2021 3D printed multi-frequency vibrational energy harvester *IEEE 20th Int. Conf. on Micro and Nanotechnology for Power Generation and Energy Conversion Applications Power MEMS*

[26] Feng Y, Zhang X, Han Y, Yu Z and Lou W 2017 Airflow-driven rotary electret energy harvester *12th Int. Conf. on Nano/Micro Engineered and Molecular Systems (NEMS) (Los Angeles, CA)*

[27] Yadav M, Yadav D, Garg R K, Gupta R K, Kumar S and Chhabra D 2021 *Advances in Fluid and Thermal Engineering* (Lecture Notes in Mechanical Engineering) (Singapore: Springer)

[28] Rafiee M, Granier F, Tao R, Bhérer-Constant A, Chenier G and Therriault D 2022 Multi-material, multi-process, planar, and nonplanar additive manufacturing of piezoelectric devices *Adv. Eng. Mater.* **24** 2200294

[29] Mahmud M *et al* 2022 Advanced design, fabrication, and applications of 3D-printable piezoelectric nanogenerators *Electron. Mater. Lett.* **18** 129–44

[30] Shepelin N A *et al* 2019 3D printing of poly(vinylidene fluoride-trifluoroethylene): a poling-free technique to manufacture flexible and transparent piezoelectric generators *MRS Commun.* **9** 159–64

[31] Rauter L *et al* 2022 Printed wireless battery-free sensor tag for health monitoring of polymer composites *IEEE Int. Conf. on Flexible and Printable Sensors and Systems (FLEPS)*

[32] Liu X, Shang Y, Zhang J and Zhang C 2021 Ionic liquid-assisted 3D printing of self-polarized β-PVDF for flexible piezoelectric energy harvesting *ACS Appl. Mater. Interfaces* **13** 14334–41

[33] Li H, Song Y S, Kim T W, Lee M H and Lim S 2022 Fully printed flexible piezoelectric nanogenerators with triethoxyvinylsilane (TEVS) coated barium titanate (BTO) nanoparticles for energy harvesting and self-powered sensing *Macromol. Mater. Eng.* **307** 2200235

[34] Rahman S, Arshad M, Qureshi A and Ullah A 2020 Fabrication of a self-healing, 3D printable, and reprocessable biobased elastomer *ACS Appl. Mater. Interfaces* **12** 51927–39

[35] Liu X, Liu J, He L, Shang Y and Zhang C 2022 3D printed piezoelectric-regulable cells with customized electromechanical response distribution for intelligent sensing *Adv. Funct. Mater.* **32** 2201274

[36] Stephen A *et al* 2022 3D-printed flexible anode for high-performance zinc ion battery *MRS Commun.* **12** 894–901

[37] Li C, Du J, Gao Y, Bu F, Tan Y H, Wang Y, Fu G, Guan C, Xu X and Huang W 2022 Stereolithography of 3D sustainable metal electrodes towards high-performance nickel iron battery *Adv. Funct. Mater.* **32** 2205317

[38] Katsuyama Y, Kudo A, Kobayashi H, Han J, Chen M, Honma I and Kaner R B 2022 A 3D-printed, freestanding carbon lattice for sodium ion batteries *Small* **18** 2202277

[39] Costa G, Lopes P A, Sanati A L, Silva A F, Freitas M C and Tavakoli M 2022 3D printed stretchable liquid gallium battery *Adv. Funct. Mater.* **32** 2113232

[40] Pinto R S, Gonçalves R, Lanceros-Méndez S and Costa C M 2022 Three-dimensional printing for solid-state batteries *Solid State Batteries 2: Materials and Advanced Devices* (American Chemical Society) pp 331–50

[41] Narita K *et al* 2022 Additive manufacturing of 3D batteries: a perspective *J. Mater. Res.* **37** 1535–46

[42] Zeng L, He H, Chen H, Luo D, He J and Zhang C 2022 3D printing architecting reservoir-integrated anode for dendrite-free, safe, and durable Zn batteries *Adv. Energy Mater.* **12** 2103708

[43] Ben-Barak I *et al* 2022 Drop-on-demand 3D-printed silicon-based anodes for lithium-ion batteries *J. Solid State Electrochem.* **26** 183–193

[44] Tian X and Xu B 2021 3D printing for solid-state energy storage *Small Methods* **5** 2100877

[45] Zhang F, Wu K, Xu X, Wu W, Hu X, Yu K and Liang C 2021 3D printing of graphite electrode for lithium-ion battery with high areal capacity *Energy Technol.* **9** 2100628

[46] Gao W and Pumera M 2021 3D printed nanocarbon frameworks for Li-ion battery cathodes *Adv. Funct. Mater.* **31** 2007285

[47] Tata Rao L, Rewatkar P, Dubey SK, Javed A and Goel S 2020 Performance optimization of microfluidic paper fuel-cell with varying cellulose fiber papers as absorbent pad *Int. J. Energy Res.* **44** 3893–904

[48] Scarcia U, Berselli G, Melchiorri C, Ghinelli M and Palli G 2016 Optimal design of 3D printed spiral torsion springs *Conf. on Smart Materials, Adaptive Structures and Intelligent Systems (Stowe, VT)* (New York: ASME) 2

[49] Miethe J F, Luebkemann F, Schlosser A, Dorfs D and Bigall N C 2020 Revealing the correlation of the electrochemical properties and the hydration of inkjet-printed CdSe/CdS semiconductor gels *Langmuir* **36** 17

IOP Publishing

3D Printed Smart Sensors and Energy Harvesting Devices
Concepts, fabrication and applications
Sanket Goel and Sohan Dudala

Chapter 12

3D-printing and sensor applications of geopolymers

R S Krishna, Suman Saha, Kurra Suresh and Amrita Priyadarshini

Geopolymers have been primarily developed as building materials but have now been established as multifunctional materials owing to their diverse applications. The introduction of filler materials in geopolymers has allowed them to develop smart properties which have manifested sensing capabilities in the matrix. 3D printable geopolymer-based self-sensors could eventually address the challenges related to structural health monitoring and other applications through conventional approaches. The integration of two separate domains, such as 3D printing and geopolymers, is considered to be a way forward for achieving sustainable development goals. This chapter presents a fundamental description of geopolymers and their application in three parts. The first part of the chapter begins with extensive details about the background of geopolymers, along with their precursor materials and preparation methodologies. The second part discusses the prevailing 3D printing techniques in the global scenario and states the essential parameters responsible for the properties of printed geopolymers, which are highly relevant for understanding the topic. Finally, the third part of the chapter highlights the properties of geopolymers for sensing applications and follows up with the developments on the current and prospective geopolymer sensors, including moisture, thermal, strain, and chloride sensors, etc.

Abbreviations

3DP	3D printing
AAS	alkali-activated slag
AM	additive manufacturing
BFS	blast furnace slag
C&D	construction and demolition
CB	carbon black

doi:10.1088/978-0-7503-5351-9ch12   12-1   © IOP Publishing Ltd 2024. All rights,

CS	compressive strength
DOF	degrees of freedom
FA	fly ash
FCA	ferrochrome ash
FCS	ferrochrome slag
FS	flexural strength
GFRP	glass fibre reinforced polymer
GGBFS	ground granulated blast furnace slag
GO	graphene oxide
GPC	geopolymer composite
ILBS	inter-layer bond strength
IOT	iron ore tailing
IS	Indian Standards
MK	metakaolin
MWCNT	multi-walled carbon nanotubes
OPC	ordinary Portland cement
rGO	reduced graphene oxide
RHA	rice husk ash
RM	red mud
SF	silica fume
SS	stainless steel
US-EPA	United States Environmental Protection Agency

12.1 Introduction

Geopolymer composites (GPCs) have been studied by various researchers across the globe for more than two decades. The first-generation GPC was initially developed by Davidovits in the 1980s and, in general, emitted fewer greenhouse gases owing to its low calcium carbonate composition-based raw materials and production temperature. GPCs are primarily developed through the chemical reaction between a highly alkaline solution and the Si–Al elements found in the binding agent [1–3]. GPCs have emerged as an eco-friendly substitution for conventional cementitious and polymeric composites. The term 'geopolymers' indicates a group of inorganic aluminosilicate polymers that are synthesized artificially in a way similar to the natural/artificial synthesis of zeolites or molecular sieves. The versatility of geopolymers allowed the use of these materials in several applications, including ceramics, coatings, 3D printing, adsorbents, self-sensing and porous materials, etc, and as a replacement for OPC in structural applications, and they have attracted considerable attention for their significant properties, such as low cost, chemical stability, corrosion and heat resistance, fast strength gain, reduced shrinkage, freeze and thaw resistance, low density, simple processing, and eco-friendliness, etc, thus garnering a lot of industrial attention [4–9]. Moreover, monumental developments have been seen in this field with several breakthrough research findings; the use of 100% industrial waste as a geopolymer precursor material is one of them, as the initial geopolymeric materials were based on naturally occurring aluminosilicates. The synergic efforts by several researchers to develop high-strength GPC at ambient temperature curing is another major accomplishment that has widened its scope for *in situ* applications and lessened the embodied energy of GPC. Similarly, continuous exploration of novel technologies is essential to promote growth and development in the field of GPC, owing to its enhanced complex material functionality and environmental benefits [10–15].

Additive manufacturing (AM) or 3D printing (3DP) is one of the most accepted applications for the production of GPC. Digital fabrication techniques hold the potential to revolutionize traditional or conventional production processes with substantial advantages to swiftly produce complicated and non-standard shapes with fine details without the absolute need for any kind of formwork [16]. Through the elimination of formwork, AM reduces 30%–65% of the cost [17] and time [17] leading to the reduction of waste incurred in the production process of GPC. Geopolymers are often considered to be moderately conductive materials due to the unique alkali ionic network microstructure. This, in turn, allows geopolymers to be utilized as sensing materials with smart features such as electrical energy storage, perceiving mechanical stress etc [18]. This study investigates the progress and developments of GPC in the field of 3DP and sensing applications while considering the relevant research gaps through the numerous research investigations published over the years.

12.2 Geopolymers

12.2.1 Precursors

Several research works have utilized different kinds of industrial wastes as binders and activators, as shown in figure 12.1(i), to produce GPC. The industrial wastes that have been taken into consideration in this section include FA, GGBFS, IOT, RM, FCA, MK, SF, FCS, and RHA, which carry unique chemical compositions of SiO_2, Al_2O_3%, as shown in figure 12.1. The amount of aluminosilicate content in these wastes varies according to their sources. Researchers have also studied the activation of the binder materials on their own or in a combination of alkali activators, including NaOH, KOH, CsOH, Na_2SiO_3, and K_2SiO_3. Different combinations of these source materials have been investigated worldwide for the preparation of geopolymer binders. FA is widely available and is a worthy candidate for geopolymerization because of its chemical composition and easy availability. However, the GPCs prepared from 100% FA have certain limitations, such as slow setting and low early-age strength gain [19, 20]. In addition, the geopolymerization of other industrial wastes is beneficial from waste management and environmental protection perspectives. Hence efforts have been undertaken worldwide to pursue the utilization of these wastes while simultaneously supplementing them with FA for enhancing geopolymer reactions.

12.2.2 Production

Geopolymer-based products are mostly accepted throughout the globe due to the flexibility of the precursor materials and the development of different matrix forms, including paste, mortar, concrete, foam, etc, for utilization in different applications. Ganesh *et al*, in their work, produced energy-efficient GGBFS-based geopolymer bricks as per IS 3495 (Part 2): 1992 [21]. A life cycle approach was adopted by

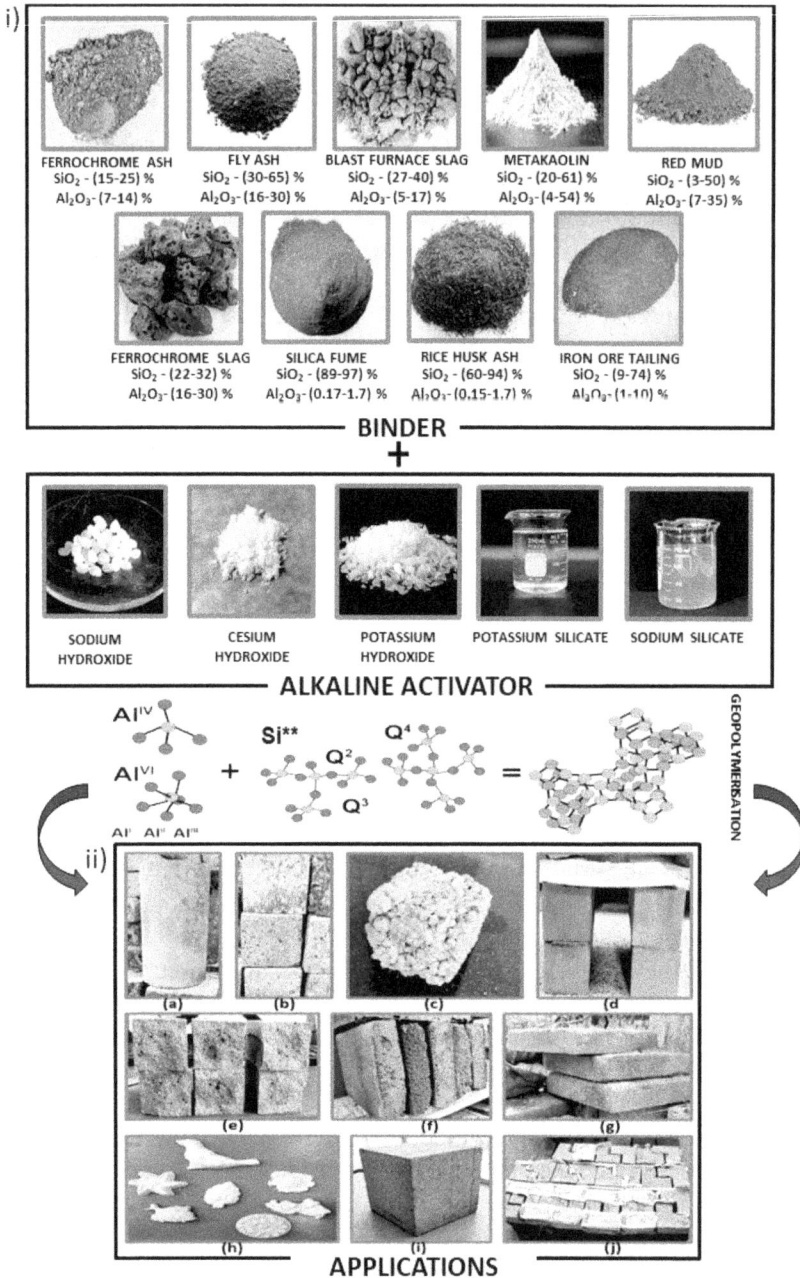

Figure 12.1. (i) Different types of binders and activators are generally utilized in geopolymer composites. (ii) Geopolymer-based products: (a) geopolymer concrete cylinders, (b) geopolymer concrete blocks, (c) geopolymer pervious concrete, (d) geopolymer sheet, (e) geopolymer beams, (f) geopolymer foams, (g) geopolymer paver blocks, (h) geopolymer casting objects, (i) lightweight aggregate concrete, (j) geopolymer buildings blocks.

Petrillo *et al* to study the impact of different parameters and carbon footprints of utilizing C&D waste in producing geopolymer paving blocks [22]. Similarly, geopolymer paving blocks were fabricated by Nath using FA and ZS while confirming the IS 15658 specifications as well as US-EPA 1311 standards. The author highlighted the importance of the immobilization of hazardous materials through geopolymer technology [23].

Further, Abdullah *et al* developed FA-based lightweight geopolymer concrete using foam technology and obtaining desired mechanical properties while stipulating the importance of curing conditions in the development process [24]. In another investigation, MK-based geopolymer tiles were prepared by Marvila *et al* to study their feasibility in high-temperature and saturation situations. Geopolymer tiles with optimum SiO_2/Al_2O_3 proved the application compatibility as per commercial requirements [25]. GFC has gained acceptance in building technology owing to its application suitability and properties. Zhang *et al* developed FA and GGBFS-based GFC using diluted aqueous surface-active concentrate as the foaming agent with acceptable CS, thermal insulation, and acoustic properties [26]. Numerous other applications are being researched and developed using geopolymer technology as it provides a sustainable means of utilizing waste materials, including structural health monitoring, 3D printing, etc. Figure 12.1(ii) illustrates a few of the prevailing applications of GPC which are being accepted commercially.

12.3 3D printing

12.3.1 Printing techniques

3D printing technology has gained significant attention due to its various advantages such as accessibility, automated processes, manufacturing flexibility, etc. The adoption of extrusion-based 3D printing technology for geopolymers in construction applications brings several advantages, which include higher build quality, reduced construction time, geometric flexibility, economic productivity, and minimized natural resource consumption. Certain challenges, including structural stability, material anisotropy and inhomogeneity, etc, need to be addressed extensively to increase the acceptability and adaptability of extrusion-based geopolymer additive manufacturing [27]. A few of the prominent extrusion-based geopolymer printing systems which are popular across the globe are the gantry system, delta system and robotic arm, as represented in figure 12.2(i). Each of these printing systems is preferred according to the field of applications and economic feasibility.

The gantry system requires a frame and a gantry for operating, to print the layers along three axes (X–Y–Z), whereas the delta system follows Cartesian coordinates to place the layers using three arms. Hence, the delta system is not limited to linear movement for the printing of extruded layers. The major challenge for the delta system is the stability of the nozzle, in particular when it is connected to a long arm for printing the layers. 3D printing can also be administered using a frame system. The robotic arm is based on a complex system, which consists of a flexible arm and nozzle to place the layers in order to develop complex structures. However, the cost of the robotic arm system is comparatively high, which restricts its acquisition to big

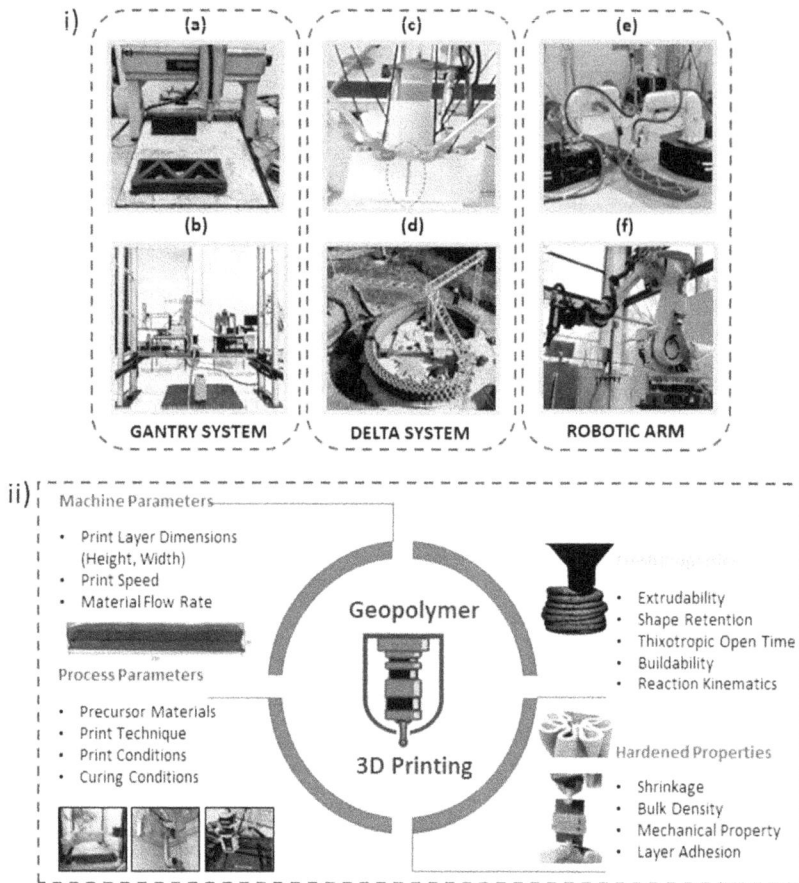

Figure 12.2. (i) Schematic illustration of different types of printing systems involved in extrusion-based geopolymer 3D printing, (ii) Schematic illustration of different process parameters and properties involved in geopolymer 3D printing. ((i)a Reproduced with permission from [49]. Copyright 2022 Elsevier. (i)b Reproduced with permission from [50]. Copyright 2022 Elsevier. (i)c Reproduced with permission from [48]. Copyright 2021 Elsevier. (i)d Reproduced with permission from [53]. Copyright 2022 Elsevier. (i)e Reproduced with permission from [52]. Copyright 2018 Elsevier. (i)f Reproduced with permission from [51]. Copyright 2020 Elsevier.)

industries or organizations. The variance between the different printing systems can be implied by their DOF, pertaining to the complexity of the printed structures. In one of the investigations, Ma *et al* developed and utilized an easy-to-handle and flexible desktop scale 3D printer with an adequate printing space of 0.7 m × 0.4 m × 0.3 m [28]. The working principles of this printer are similar to the gantry system printing technique. Cao *et al*, in their extensive review, explained the different techniques for 3D printing systems in detail and compared each technique, and concluded that gantry/frame systems are suitable for beginners, whereas robot arm systems are suitable for research/industrial works as they accommodate the freedom to print with different axes. Delta systems were not recommended for printing

Table 12.1. Specifications of the globally accepted construction 3D printers.

Company	Printing system	Printer	Build size (m)	Country	References
BetAbram	Gantry	P1	16 × 8.2 × 2.5	Slovenia	[30]
COBOD	Gantry	BOD2	14.62 × 50.52 × 8.14	Denmark	[31]
Constructions-3D	Robotic arm	MAXI PRINTER	12.25 × 12.25 × 7	France	[32]
CyBe	Robotic arm	RC 3Dp	2.75 × 2.75 × 2.75	Netherlands	[33]
ICON	Gantry	Vulcan II	2.6 × 8.5 × ∞	USA	[34]
MudBots	Gantry	Concrete 3D printer	1.83 × 1.83 × 1.22	USA	[35]
Total Kustom	Gantry	StroyBot 6.2	10 × 20 × 6	USA	[36]
WASP	Dela	Infinity 3D printer	Ø 6.3 × 3	Italy	[37]
Apis Cor	Robotic arm	3D printer	8.5 × 1.6 × 1.5	USA	[38]
Batiprint3D	Robotic arm	3D printer	7 (height)	Italy	[39]
SQ4D	Gantry	ARCS	9.1 × 4.4 × ∞	USA	[40]
XtreeE	Robotic arm	3D printer	—	France	[41]

applications due to their higher operating process and complexity [29]. A few of the notable organizations incorporating the discussed printing systems are included in table 12.1.

For large-scale applications, the 3D printing techniques are considered based on different aspects such as availability, requirements, maintenance, economic viability, etc. In addition, the time intervals between two subsequent layers, exposure and curing conditions, nozzle standoff distance, and nozzle variations are identified as crucial parameters for extrusion-based 3D printing, which affect the inter-layer properties of the 3D-printed structure [42–45]. Review studies with the consideration of all such parameters involving different precursor materials for specific printing techniques are limited and are recommended for future studies to obtain the complete scope of geopolymer 3D printing.

12.3.2 Process parameters and properties

In this chapter the primary emphasis is placed on the utilization of geopolymer materials for 3D printing in structural applications. Standardization of geopolymer materials at a global scale has been one of the major restraints, which has posed difficulties in the adjustment and forecasting of the applicable process parameters and properties, both for conventional as well as additive manufacturing. The variation in the preference of appropriate source/precursor materials and activating solutions (as illustrated in figure 12.1(i)), aggregates, rheological/functional additives, reinforcing elements, along with mix proportioning and curing regimes, which differ from country to country, add to the complexity of the appropriate optimization process for manufacturing of geopolymer paste/mortar/concrete [46]. Fly ash and GGBFS have remained the most popular precursor materials for the 3D printing of geopolymers due to their surplus availability and convenient mixture proportioning aspects for obtaining superior properties. An overview of the essential parameters and properties associated with geopolymer 3D printing is shown in figure 12.2(ii). In addition, certain internal and external parameters also produce positive effects on the mechanical strength of the 3D-printed geopolymers. The internal parameters include precursor material (type, proportion, activity), liquid-to-solid ratio, and activator (type, concentration, Si/(Na or K)), and the external parameters consist of temperature, curing system (curing medium, ambient/microwave heating, curing period), and humidity [47]. In order to obtain an effective geopolymer printable mix, skillful tailoring of all of the critical parameters must be taken into consideration. Several research investigations have focused on the influence of machine and process parameters on the fresh and hardened properties of the 3D-printed geopolymers, and have been considered to investigate and comprehend the interrelationship among them.

Different researchers have adopted distinctive ways to investigate and measure the interdependence between the adopted parameters and the resulting properties. Therefore, it is difficult to determine an optimum range of the parameters and the properties based on the recommended values of the published results. However, these values provide an overall approximation to achieve favourable 3D-printed geopolymer structures. The range of values for the input parameters include: print speed (18–130 mm s^{-1}), pipe length (2–5 m), pipe diameter (20–30 mm), pump flow rate (0.5–7 l min^{-1}), gantry speed (50–90 mm s^{-1}), nozzle size (300–500 mm^2, 1–30 mm), time interval between layers (2–20 min), layer height/thickness (3–20 mm), extrusion rate (40%–60%), print time (15–45 min), pump pressure (8–18 bar), liquid-to-solid ratio (0.18–0.95), etc [46, 47, 54]. Similarly, the mechanical properties of the hardened 3D-printed geopolymer elements range as CS (8–80 MPa), FS (6–20 MPa), and ILBS (1.5–10).

Table 12.2 provides an overview of the research progress on extrusion-based 3D printing using geopolymer mixes in all forms (paste/mortar/concrete). It can be observed from table 12.2 that extrusion-based printing techniques are most often used compared to powder-based techniques, and research studies primarily concentrate on geopolymer paste or mortar matrix forms to obtain superior end properties. The nozzle dimensions play a significant role in order decide the maximum size in the printable concrete mixes. Therefore, there is a need for more scientific studies on the printable mix in concrete form. Comparatively fewer studies have been conducted with the incorporation of reinforcing elements such as fibres, nano-materials, etc, to enhance the rheological and mechanical properties of the geopolymer mix [55–57]. Hence, greater attention is required to design a geopolymer printable mix with the consideration of all parameters in order to have the desired fresh, hardened properties of geopolymer concrete while achieving economic

Table 12.2. Research progress in the field of geopolymer 3D printing.

Matrix	Binder	Activator	Curing conditions	3D printing type	Filler (Conc.)	Properties	Application	Ref
Paste	MK+FA	KOH (15M) + K$_2$SiO$_3$	48–72 h (ambient temperature)	Extrusion	—	Compressive strength: 8.5 MPa, Bulk density: 1.30 g cm^{-3} Apparent density: 2.23 g cm^{-3} True density: 2.436 g cm^{-3} Total porosity: 71%	Sacrificial templates	[58]
Mortar	Slag powder	Na$_2$SiO$_3$	2 h (ambient temperature) + 7 days (60 °C, Na$_2$SiO$_3$)	Powder based	—	Compressive strength: 16.5 MPa Bulk density: 1.20 g cm^{-3} True density: 2.81 g cm^{-3} Apparent porosity: 57.1%,	Construction	[5]
Paste	ASOPs	ALSPs	17 °C (5 days) + 1000 °C (0.5 h)	Extrusion	GO (5 wt%)	Compressive strength: 36 MPa Young's modulus: 700 MPa Electrical conductivity: 10^2 S m^{-1}	Construction	[59]
Mortar	FA (Class F) + GGBFS+ MS	NaOH (8M) + Na$_2$SiO$_3$	28 days (ambient temperature)	Extrusion	—	Compressive strength: 36.25 MPa Flexural strength: 9.5 MPa Bulk density: 2250 kg m^{-3}	Construction	[60]
Mortar	FA (Class F), GGBFS + MS	K$_2$SiO$_3$	28 days (ambient temperature)	Extrusion	Glass fibre (1 wt%)	Compressive strength: 26.5 MPa Flexural strength: 7 MPa	Construction	[55]
Mortar	FA (Class F)	NaOH (8M), Na$_2$SiO$_3$	60 °C (24 h)	Extrusion	Polypropylene fibres (0.25 vol%)	Compressive strength: 36 MPa Flexural strength: 7.8 MPa Apparent porosity: 10.1% Inter-layer bond strength: 3.1 MPa	Construction	[56]

(*Continued*)

Table 12.2. (*Continued*)

Matrix	Binder	Activator	Curing conditions	3D printing type	Filler (Conc.)	Properties	Application	Ref
Concrete	FA (Class F)	NaOH (8M), Na₂SiO₃	60 °C (24 h) + 2 days (ambient temperature)	Extrusion	Polyphenylene benzobi soxazole	Flexural strength: 10.3 MPa Inter-layer bond strength: 2.33 MPa	Construction	[61]
Concrete	FA (Class F), GGBFS	Na₂SiO₃	60 °C (24 h)	Extrusion	—	Compressive strength: 25.2 MPa Flexural strength: 6.1 MPa Inter-layer bond strength: 1.3 MPa	Construction	[62]
Mortar	FA (Class F), GGBFS + MS	K₂SiO₃	7 days (ambient temperature)	Extrusion	Polyvinyl alcohol (PVA) (0.5 wt %) + stainless steel cable (SUS304– 2 mm)	Flexural strength: 290% (increase)	Construction	[63]
Mortar	Slag-based geopolymer powder	Na₂SiO₃	6 h (ambient conditions) + 60 °C (7 days)	Powder based	—	Compressive strength: 19.3 MPa	Construction	[64]
Paste	FA (Class F)	NaOH (8M), Na₂SiO₃	6 h (ambient conditions) + 60 °C (7 days)	Powder based	—	Compressive strength: 29.6 MPa	Construction	[65]
Mortar	FA (Class F)	NaOH (10M), Na₂SiO₃	75 °C (24 h) + 28 days (ambient temperature)		Flax fibre (1 wt%)	Compressive strength: 48.7 MPa Flexural strength: 9.4 MPa	Construction	[57]

diversity. In addition to the construction industry, additive manufacturing of geopolymers also provides opportunities for various advanced energy and environmental applications such as catalysts, water purification filters, conductive materials, smart sensors, etc [46].

12.4 Self-sensing applications

Geopolymers on a micro level are mostly amorphous materials which consist of cross-linked alumane and tetrahedral silicate chains with unchained alkali ions acting as charge carriers, permitting geopolymers to act as ionic conductors [66]. Variations in the ionic conductivity enable geopolymers to be utilized as sensors [8, 67]. Therefore, intensive research can be seen on the fabrication of building materials/geopolymer matrices for potential application in a broad range of sensors.

A subtle approach was adopted by Rovnaník *et al* to characterize the electrical properties of BFS-based AAS under repeated compressive loading of 5 MPa from two brass electrode surfaces, as illustrated in figure 12.3(a) [68]. The test results revealed an exceptional stress sensitivity coefficient of 97.13 Ω MPa^{-1}, which mostly contributed due to the combination of ionic, contact, and tunnelling conduction of iron particles from BFS, consequently aiding self-sensing applications without the need for conductive fillers. In a study by McAlorum *et al*, the authors employed an unconventional approach to investigate the smart/self-sensing properties of metakaolin-based geopolymers through robotic spray coating, as shown in figure 12.3(b) for concrete repair applications [69]. The robotic spray process parameters, such as spray pressure, pump volume flow, nozzle diameter, etc, were optimized to a laboratory scale to obtain a bond strength of 0.5 MPa with the concrete surface. The geopolymer coating obtained the strain and temperature measurements of the concrete surface with resolutions of 1 $\mu\varepsilon$ and 0.2 °C through a set of four embedded metal electrodes. Further, figure 12.3(c) highlights the application of 3D-printed alkali-activated material patches as strain sensors for concrete as well as GFRP surfaces, whereas figure 12.3(d) shows the applicability of 3D-printed geopolymer patches as temperature sensors on concrete substrates [70, 71]. Feasibility studies were carried out by Vlachakis *et al* through the transitions in electrical impedance parameters. For figure 12.3(c) the obtained gauge factors were 8.6 under compression (concrete) and 38.4 under tension (GFRP), and for figure 12.3(d) the temperature sensing precision was 0.1 °C with a repeatability of 0.3 °C. Another investigation by Perry *et al* highlighted a novel geopolymer patch compromising ground magnetite (0–60 wt%) for temperature sensing applications. Figure 12.3(e) illustrates the novel geopolymer patch embedded with electrical probes and an induction coil for heating the patch [9]. Different thermal signatures through impedance decay were recorded for different environments indicating the presence of air, water, soil, etc, demonstrating the feasibility in applications such as scouring wind turbine foundations, deep sea cables and bridges.

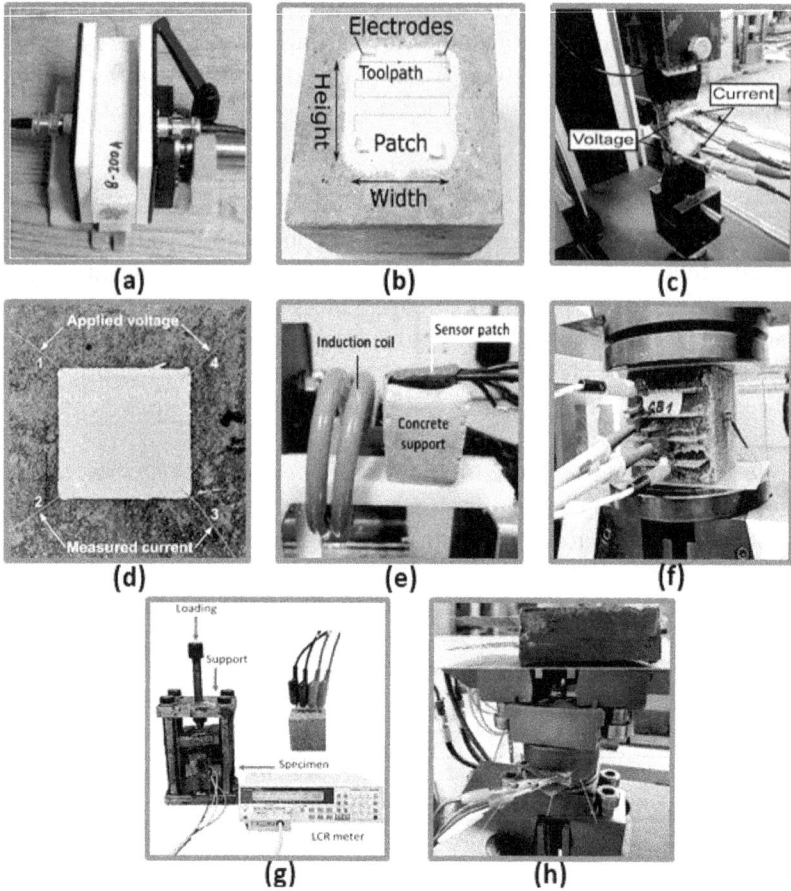

Figure 12.3. (a) Geopolymer mortar, (b) geopolymer spray coating, (c) geopolymer patch on GFRP, (d) 3D-printed geopolymer patch, (e) geopolymer magnetite patch, (f) geopolymer CB mortar, (g) geopolymer MWCNT mortar, and (h) geopolymer paste. ((a) Reproduced with permission from [69]. Copyright 2019 Elsevier. (b) Reproduced with permission from [70]. Copyright 2021 Elsevier. (c) CC-BY 4.0. (d) Reproduced with permission from [72]. Copyright 2020 Elsevier. (e) CC0. (f) CC-BY 4.0. (g) Reproduced with permission from [74]. Copyright 2021 Elsevier. (h) Reproduced with permission from [18]. Copyright 2018 Elsevier.)

The introduction of conductive fillers for the enhancement of electrical properties in building materials has been one of the apparent choices of a number of researchers. In an observation, Mizerová *et al* analysed the piezoresistivity of the FA-based geopolymers doped with CB (0.5 wt%) under cyclic compression loading. The gauge factor and the initial resistance were recorded to be 6 and 33.6 Ω, respectively, through the copper electrodes embedded in the geopolymer matrix, as depicted in figure 12.3(f) [72]. The stiffness of the geopolymer matrix was also observed to decrease with the increase in CB dosages while altering the state of the material from quasi-brittle to ductile. Similarly, MWCNTs were added by Maho *et al* to evaluate the electrical resistivity of high calcium FA-based geopolymer mortars for piezoelectric sensor applications [73]. Figure 12.3(g) represents the evaluation of the resistivity parameter under varying loading conditions which resulted in increased resistivity within 10% of the ultimate load and vice versa. Electrical energy storage in structural components has been one of the most popular fields of investigation due to the potential to be utilized as structural capacitors for aiding sensing and monitoring structural integrity applications. Saafi *et al* fabricated fly ash-based and potassium silicate-activated geopolymers with steel mesh electrodes, as shown in figure 12.3(h), to determine the conduction and energy storage parameters [18]. The ionic conductivity of the geopolymers ranged between $12 (10^{-2} \, S \, m^{-1})$ with a maximum activation energy of 0.97 eV and power density of $0.33 \, kW \, m^{-2}$ with 2 h of discharge life.

The above studies have highlighted the feasibility of geopolymers in sensing applications, although on a laboratory scale. However, certain parameters have higher significance in affecting the conductivity (i.e. including electrical impedance) of geopolymeric composites, which include temperature, moisture, ionic contamination, strain and cracking. Research progress related to the development of geopolymer sensors and the associated specifications is detailed in table 12.3.

12.5 Summary and conclusions

Increased global sustainability demands have allowed the growth and acceptance of geopolymers in the construction and materials industry. This chapter provides a brief synopsis regarding the 3D printing of geopolymers and the application of geopolymers with regard to self-sensing applications. The advanced 3D printing extrusion-based techniques utilized for geopolymers and their related parameters and properties are summarized and discussed. A detailed review of the current research progress of geopolymers for sensor applications has been carried out, which indicates the specific needs in the construction sector. However, most of the research outputs are on a laboratory scale. Therefore, there is an immense need to substantiate the feasibility of these applications with the necessary performance on an industrial and commercial scale.

Acknowledgments

The authors are grateful for the academic support from the Birla Institute of Technology and Science Pilani (BITS-Pilani), Hyderabad Campus and National Institute of Technology Durgapur, West Bengal, India.

Table 12.3. Research progress in the field of geopolymer sensor applications.

Matrix	Binder	Activator	Mix proportion	Filler (conc.)	Assessment	Results	Application	Ref
Mortar	Metakaolin	NaOH + Na$_2$SiO$_3$	SiO$_2$/Al$_2$O$_3$: (3.8–6)	No filler	Optical fibre	Tensile cracking strains: 0.08%–0.16%	Strain sensing	[74]
Concrete	Fly ash	NaOH (10M) + Na$_2$SiO$_3$	L/B: 0.5 Na$_2$SiO$_3$/NaOH: 2.5	Carbon fibre (0.4 wt%)	24 gage copper wire (wounded), silver paste	Resistivity: 157.91 Ω	Strain sensing	[75]
Paste	Fly ash	NaOH (10M) + Na$_2$SiO$_3$	L/B: 0.389 Na$_2$SiO$_3$/NaOH: 2.5	rGO (0.35 wt%)	Copper electrodes (inserted)	Conductivity: 2.38 S m^{-1}; Gauge factor: 20.7 (tension); Gauge factor: 43.87 (compression)	Self-sensing	[76]
Mortar	Metakaolin	NaOH + Na$_2$SiO$_3$	Si/Al: 1.75	No Filler	Fibre Bragg gratings (embedded)		Temperature and shrinkage sensor	[77]
Paste	Fly ash	NaOH (10M) + Na$_2$SiO$_3$	L/B: 0.389 Na$_2$SiO$_3$/NaOH: 2.5	Magnetite (35 wt%)	Electrical probes (embedded)	Thermal decay signature: 1–4 W m^{-1} K	Thermal conductivity sensor	[9]
Paste	Fly ash	—	—	No filler	Graphene electrodes (embedded)	Conductivity: (1.54–1.72) 10^{-2} S m^{-1}; Activation energy: 0.156 eV; Temperature sensitivity: 21.5 kΩ °C^{-1}; Gauge factor: 358 (tension)	Superionic long gauge sensor	[78]
Paste	Metakaolin	NaOH + Na$_2$SiO$_3$	Si:Al: 1.9:1	SiO$_2$ coated CNT (0.25 vol%)	Copper wire electrodes (embedded)	Gauge factor: 663.3 (compression); Gauge factor: 724.6 (flexural)	Structural sensor	[79]

Matrix	Precursor	Activator	Ratio	Filler	Electrodes	Properties	Application	Ref.
Paste	Fly ash	$KOH + Na_2SiO_3$	Na_2SiO_3/KOH: 2.34	Graphene (1 wt %)	Electrodes (inserted)	Conductivity: 0.3 S m^{-1}	Photocatalytic sensor	[80]
Paste	Fly ash	K_2SiO_3	L/B: 0.6	No filler	Steel mesh electrodes (Inserted)	Conductivity: 12 (10-2 S m^{-1}); Activation energy: 0.97 eV; Power density: 0.33 kW m^{-2}; Discharge life: 2 h; Compression sensor: 11 Ω MPa^{-1}, 0.55° MPa^{-1}	Energy storage, structural sensor	[18]
Paste	Fly ash	$NaOH + Na_2SiO_3$	L/B: 0.5; $Na_2SiO_3/NaOH$—2.5	No filler	Seven-wire braided SS wire (embedded)	Moisture precision: (0.02–0.26 wt%); Temperature precision: (0.06 °C–0.18 °C)	Moisture and temperature sensor	[8]
Paste	Metakaolin	H_3PO_4 (10 M)	—	No filler	Carbon paste electrode	Detection limit: 2.3 × 10^{-9} (mol l^{-1}); Sensitivity: 52.17 (µA µM^{-1}); Linearity domain: [0.1–1.4 (µmol l^{-1})]	Electrochemical sensor	[81]
Mortar	Fly ash	NaOH (12 M) + Na_2SiO_3	L/B: 0.45; $Na_2SiO_3/NaOH$: 1	MWCNT (0.6 wt%)	Copper sheet electrodes (inserted)	Electrical resistivity: [2.2–3.2 (Ω^{-m})]	Piezoelectric structural sensor	[73]
Mortar	Fly ash	Na_2SiO_3	—	Graphite powders (10 wt%)	Copper electrodes (inserted)	Resistance: 0.134 k Ω	Structural sensor	[82]

References

[1] Davidovits J and Davidovits R 2020 Ferro-sialate geopolymers (–Fe–O–Si–O–Al–O–) *Technical Paper* #27 Geopolymer Institute Library

[2] Davidovits J 1991 Geopolymers—inorganic polymeric new materials *J. Therm. Anal.* **37** 1633–56

[3] Krishna R S, Mishra J, Zribi M, Adeniyi F, Saha S, Baklouti S, Shaikh F U A and Gökçe H S 2021 A review on developments of environmentally friendly geopolymer technology *Materialia* **20** 101212

[4] Yu H, Xu M, Chen C, He Y and Cui X 2022 A review on the porous geopolymer preparation for structural and functional materials applications *Int. J. Appl. Ceram. Technol.* **19** 1793–813

[5] Xia M and Sanjayan J 2016 Method of formulating geopolymer for 3D printing for construction applications *Mater. Des.* **110** 382–90

[6] Aguirre-Guerrero A M, Robayo-Salazar R A and de Gutiérrez R M 2017 A novel geopolymer application: coatings to protect reinforced concrete against corrosion *Appl. Clay Sci.* **135** 437–46

[7] Liew Y-M, Heah C-Y, Li L, Jaya N A, Abdullah M M A B, Tan S J and Hussin K 2017 Formation of one-part-mixing geopolymers and geopolymer ceramics from geopolymer powder *Constr. Build. Mater.* **156** 9–18

[8] Biondi L, Perry M, McAlorum J, Vlachakis C and Hamilton A 2020 Geopolymer-based moisture sensors for reinforced concrete health monitoring *Sens. Actuators* B **309** 127775

[9] Perry M, Saafi M, Fusiek G and Niewczas P 2016 Geopolymeric thermal conductivity sensors for surface-mounting onto concrete structures *9th Int. Concrete Conf. (UK)*

[10] Krishna R S, Mishra J, Meher S, Das S K, Mustakim S M and Singh S K 2020 Industrial solid waste management through sustainable green technology: case study insights from steel and mining industry in Keonjhar, India *Mater. Today Proc.* **33** 5243–9

[11] Mishra J, Kumar Das S, Krishna R S, Nanda B, Kumar Patro S and Mohammed Mustakim S 2020 Synthesis and characterization of a new class of geopolymer binder utilizing ferrochrome ash (FCA) for sustainable industrial waste management *Mater. Today Proc.* **33** 5001–6

[12] Mishra J, Das S K, Krishna R S and Nanda B 2020 Utilization of ferrochrome ash as a source material for production of geopolymer concrete for a cleaner sustainable environment *Indian Concr. J.* **94** 40–9

[13] Krishna R S, Shaikh F, Mishra J, Lazorenko G and Kasprzhitskii A 2021 Mine tailings-based geopolymers: properties, applications and industrial prospects *Ceram. Int.* **47** 17826–43

[14] Lazorenko G, Kasprzhitskii A, Shaikh F, Krishna R S and Mishra J 2021 Utilization potential of mine tailings in geopolymers: physicochemical and environmental aspects *Process Saf. Environ. Prot.* **147** 559–77

[15] Das S K, Krishna R S, Mishra S, Mustakim S M, Jena M K, Tripathy A K and Sahu T 2021 Future trends nanomaterials in alkali-activated composites *Handbook of Sustainable Concrete and Industrial Waste Management* ed F Colangelo, R Cioffi and I Farina (Amsterdam: Elsevier)

[16] Lipson H and Kurman M 2013 *Fabricated: The New World of 3D Printing* (New York: Wiley)

[17] Lloret E, Shahab A R, Linus M, Flatt R J, Gramazio F, Kohler M and Langenberg S 2015 Complex concrete structures: merging existing casting techniques with digital fabrication *Comput. Aided Des.* **60** 40–9

[18] Saafi M, Gullane A, Huang B, Sadeghi H, Ye J and Sadeghi F 2018 Inherently multifunctional geopolymeric cementitious composite as electrical energy storage and self-sensing structural material *Compos. Struct.* **201** 766–78

[19] Huseien G F, Sam A R M and Alyousef R 2021 Texture, morphology and strength performance of self-compacting alkali-activated concrete: role of fly ash as GBFS replacement *Constr. Build. Mater.* **270** 121368

[20] Tennakoon C, Nicolas R S, Sanjayan J G and Shayan A 2016 Thermal effects of activators on the setting time and rate of workability loss of geopolymers *Ceram. Int.* **42** 19257–68

[21] Chithambar Ganesh A, Muthukannan M, Aakassh S, Prasad and Subramanaian B 2020 Energy efficient production of geopolymer bricks using industrial waste *IOP Conf. Ser.: Mater. Sci. Eng.* **872** 012154

[22] Petrillo A, Cioffi R, Ferone C, Colangelo F and Borrelli C 2016 Eco-sustainable geopolymer concrete blocks production process *Agric. Agric. Sci. Procedia* **8** 408–18

[23] Nath S K 2020 Fly ash and zinc slag blended geopolymer: Immobilization of hazardous materials and development of paving blocks *J. Hazard. Mater.* **387** 121673

[24] Abdullah M M A B, Hussin K, Bnhussain M, Ismail K N, Yahya Z and Abdul Razak R 2012 Fly ash-based geopolymer lightweight concrete using foaming agent *Int. J. Mol. Sci.* **13** 7186–98

[25] Marvila M T, Azevedo A R G, Delaqua G C G, Mendes B C, Pedroti L G and Vieira C M F 2021 Performance of geopolymer tiles in high temperature and saturation conditions *Constr. Build. Mater.* **286** 122994

[26] Zhang Z, Provis J L, Reid A and Wang H 2015 Mechanical, thermal insulation, thermal resistance and acoustic absorption properties of geopolymer foam concrete *Cem. Concr. Compos.* **62** 97–105

[27] Qaidi S, Yahia A, Tayeh B A, Unis H, Faraj R and Mohammed A 2022 3D printed geopolymer composites: a review *Mater. Today Sustain.* **20** 100240

[28] Ma G, Li Y, Wang L, Zhang J and Li Z 2020 Real-time quantification of fresh and hardened mechanical property for 3D printing material by intellectualization with piezoelectric transducers *Constr. Build. Mater.* **241** 117982

[29] Cao X, Yu S, Cui H and Li Z 2022 3D printing devices and reinforcing techniques for extruded cement-based materials: a review *Buildings* **12** 453

[30] Anon 2022 BetAbram P1 overview *Aniwaa* https://www.aniwaa.com/product/3d-printers/betabram-p1/

[31] Anon 2022 The Bod2 *Cobod* https://cobod.com/products/bod2/

[32] Anon 2022 MaxiPrinter *Constructions-3D* https://en.constructions-3d.com/la-maxi-printer

[33] Anon 2022 CyBe RC (Robot Crawler) *CyBe Construction* https://cybe.eu/3d-concrete-printing/printers/

[34] Anon 2022 ICON vulcan II *Aniwaa* https://www.aniwaa.com/product/3d-printers/icon-vulcan-ii/

[35] Anon 2022 3D concrete printers *MUDBOTS 3D CONCRETE PRINTING* https://www.mudbots.com/concrete-3d-printers.php

[36] Anon 2022 StroyBot 6.2 *Total Kustom* http://www.totalkustom.com/3d-concrete-printers.html

[37] Anon 2022 Crane WASP *WASP Srl* https://www.3dwasp.com/stampante-3d-per-case-crane-wasp/

[38] Anon 2022 Technology *Apis Cor Inc.* https://apis-cor.com/technologies/

[39] Anon 2022 Shaping tomorrow *BATIPRINT 3D* https://www.batiprint3d.com/en

[40] Anon 2022 3D printed houses, commercial buidings, infrastructure *SQ4D* https://www.sq4d.com/
[41] Anon 2022 The large-scale 3D *XtreeE* https://xtreee.com/en/solutions/
[42] Wolfs R J M, Bos F P and Salet T A M 2019 Hardened properties of 3D printed concrete: the influence of process parameters on interlayer adhesion *Cem. Concr. Res.* **119** 132–40
[43] Chen Y, Jansen K, Zhang H, Romero Rodriguez C, Gan Y, Çopuroğlu O and Schlangen E 2020 Effect of printing parameters on interlayer bond strength of 3D printed limestone-calcined clay-based cementitious materials: an experimental and numerical study *Constr. Build. Mater.* **262** 120094
[44] Tay Y W D, Ting G H A, Qian Y, Panda B, He L and Tan M J 2019 Time gap effect on bond strength of 3D-printed concrete *Virtual Phys. Prototyp.* **14** 104–13
[45] Chen Y, He S, Gan Y, Çopuroğlu O, Veer F and Schlangen E 2022 A review of printing strategies, sustainable cementitious materials and characterization methods in the context of extrusion based 3D concrete printing *J. Build. Eng.* **45** 103599
[46] Lazorenko G and Kasprzhitskii A 2022 Geopolymer additive manufacturing: a review *Addit. Manuf.* **55** 102782
[47] Raza M H, Zhong R Y and Khan M 2022 Recent advances and productivity analysis of 3D printed geopolymers *Addit. Manuf.* **52** 102685
[48] Tramontin Souza M, Simão L, Guzi de Moraes E, Senff L, de Castro Pessôa J R, Ribeiro M J and Novaes de Oliveira A P 2021 Role of temperature in 3D printed geopolymers: evaluating rheology and buildability *Mater. Lett.* **293** 129680
[49] Ilcan H, Sahin O, Kul A, Yildirim G and Sahmaran M 2022 Rheological properties and compressive strength of construction and demolition waste-based geopolymer mortars for 3D-printing *Constr. Build. Mater.* **328** 127114
[50] Bong S H, Nematollahi B, Xia M, Ghaffar S H, Pan J and Dai J G 2022 Properties of additively manufactured geopolymer incorporating mineral wollastonite microfibers *Constr. Build. Mater.* **331** 127282
[51] Comminal R, Leal da Silva W R, Andersen T J, Stang H and Spangenberg J 2020 Modelling of 3D concrete printing based on computational fluid dynamics *Cem. Concr. Res.* **138** 106256
[52] Zhang X, Li M, Lim J H, Weng Y, Tay Y W D, Pham H and Pham Q-C 2018 Large-scale 3D printing by a team of mobile robots *Autom. Constr.* **95** 98–106
[53] Bello N D and Memari A M 2022 Comparative review of the technology and case studies of 3D concrete printing of buildings by several companies *Buildings* **13** 106
[54] Zhong H and Zhang M 2022 3D printing geopolymers: a review *Cem. Concr. Compos.* **128** 104455
[55] Panda B, Chandra Paul S and Jen Tan M 2017 Anisotropic mechanical performance of 3D printed fiber reinforced sustainable construction material *Mater. Lett.* **209** 146–9
[56] Nematollahi B, Vijay P, Sanjayan J, Nazari A, Xia M, Nerella V N and Mechtcherine V 2018 Effect of polypropylene fibre addition on properties of geopolymers made by 3D printing for digital construction *Materials* **11** 2352
[57] Korniejenko K, Kejzlar P and Louda P 2022 The influence of the material structure on the mechanical properties of geopolymer composites reinforced with short fibers obtained with additive technologies *Int. J. Mol. Sci.* **23** 2023
[58] Franchin G and Colombo P 2015 Porous geopolymer components through inverse replica of 3D printed sacrificial templates *J. Ceram. Sci. Technol.* **6** 105–12
[59] Zhong J, Zhou G X, He P G, Yang Z H and Jia D C 2017 3D printing strong and conductive geo-polymer nanocomposite structures modified by graphene oxide *Carbon* **117** 421–6

[60] Panda B, Paul S C, Hui L J, Tay Y W D and Tan M J 2017 Additive manufacturing of geopolymer for sustainable built environment *J. Clean. Prod.* **167** 281–8

[61] Nematollahi B, Xia M, Sanjayan J and Vijay P 2018 Effect of type of fiber on inter-layer bond and flexural strengths of extrusion-based 3D printed geopolymer *Mater. Sci. Forum* **939** 155–62

[62] Nematollahi B, Xia M, Bong S H and Sanjayan J 2019 Hardened properties of 3D printable 'one-part' geopolymer for construction applications *First RILEM International Conference on Concrete and Digital Fabrication – Digital Concrete* RILEM Bookseries vol 19 ed T Wangler and R Flatt (Dordrecht: Springer) pp 190–9

[63] Lim J H, Panda B and Pham Q C 2018 Improving flexural characteristics of 3D printed geopolymer composites with in-process steel cable reinforcement *Constr. Build. Mater.* **178** 32–41

[64] Xia M, Nematollahi B and Sanjayan J 2018 Influence of binder saturation level on compressive strength and dimensional accuracy of powder-based 3D printed geopolymer *Mater. Sci. Forum* **939** 177–83

[65] Xia M and Sanjayan J G 2018 Methods of enhancing strength of geopolymer produced from powder-based 3D printing process *Mater. Lett.* **227** 281–3

[66] Cui X-M, Zheng G-J, Han Y-C, Su F and Zhou J 2008 A study on electrical conductivity of chemosynthetic Al_2O_3–$2SiO_2$ geopolymer materials *J. Power Sources* **184** 652–6

[67] Krishna R S, Saha S, Korniejenko K, Qureshi T S and Mustakim S M 2023 Investigation of the electrical properties of graphene-reinforced geopolymer composites *10th MATBUD'2023 Scientific-Technical Conference* (Basel: MDPI) p 34

[68] Rovnaník P, Kusák I, Bayer P, Schmid P and Fiala L 2019 Comparison of electrical and self-sensing properties of Portland cement and alkali-activated slag mortars *Cem. Concr. Res.* **118** 84–91

[69] McAlorum J, Perry M, Vlachakis C, Biondi L and Lavoie B 2021 Robotic spray coating of self-sensing metakaolin geopolymer for concrete monitoring *Autom. Constr.* **121** 103415

[70] Vlachakis C, McAlorum J and Perry M 2022 3D printed cement-based repairs and strain sensors *Autom. Constr.* **137** 104202

[71] Vlachakis C, Perry M, Biondi L and McAlorum J 2020 3D printed temperature-sensing repairs for concrete structures *Addit. Manuf.* **34** 101238

[72] Mizerová C, Kusák I, Topolář L, Schmid P and Rovnaník P 2021 Self-sensing properties of fly ash geopolymer doped with carbon black under compression *Materials* **14** 4350

[73] Maho B, Sukontasukkul P, Sua-Iam G, Sappakittipakorn M, Intarabut D, Suksiripattanapong C, Chindaprasirt P and Limkatanyu S 2021 Mechanical properties and electrical resistivity of multiwall carbon nanotubes incorporated into high calcium fly ash geopolymer *Case Stud. Constr. Mater.* **15** e00785

[74] He J, Zhang G, Asce M, Shuang H and Cai C S 2011 Geopolymer-based smart adhesives for infrastructure health monitoring: concept and feasibility *J. Mater. Civ. Eng.* **23** 100–9

[75] Vaidya S and Allouche E N 2011 Strain sensing of carbon fiber reinforced geopolymer concrete *Mater. Struct.* **44** 1467–75

[76] Saafi M, Tang L, Fung J, Rahman M, Sillars F, Liggat J and Zhou X 2014 Graphene/fly ash geopolymeric composites as self-sensing structural materials *Smart Mater. Struct.* **23** 065006

[77] Campopiano S, Iadicicco A, Messina F, Ferone C and Cioffi R 2015 Fiber Bragg grating sensors as a tool to evaluate the influence of filler on shrinkage of geopolymer matrices *Proc. SPIE* **9506** 95061J

[78] Saafi M, Piukovics G and Ye J 2016 Hybrid graphene/geopolymeric cement as a superionic conductor for structural health monitoring applications *Smart Mater. Struct.* **25** 105018

[79] Bi S, Liu M, Shen J, Hu X M and Zhang L 2017 Ultrahigh self-sensing performance of geopolymer nanocomposites via unique interface engineering *ACS Appl. Mater. Interfaces* **9** 12851–8

[80] Zhang Y J, He P Y, Zhang Y X and Chen H 2018 A novel electroconductive graphene/fly ash-based geopolymer composite and its photocatalytic performance *Chem. Eng. J.* **334** 2459–66

[81] Pengou M, Bertrand G, Ngassa P, Boutianala M, Kouamo Tchakouté H, Péguy Nanseu-Njiki C and Ngameni E 2021 Geopolymer cement-modified carbon paste electrode: application to electroanalysis of traces of lead(II) ions in aqueous solution *J. Solid State Electrochem.* **25** 1183–95

[82] Rovnaník P, Kusák I and Schmid P 2021 Self-sensing properties of fly ash geopolymer composite with graphite filler *AIP Conf. Proc.* **2322** 020016

IOP Publishing

3D Printed Smart Sensors and Energy Harvesting Devices
Concepts, fabrication and applications
Sanket Goel and Sohan Dudala

Chapter 13

4D printing: fundamentals and applications

Ayan Bhatnagar, T Lachana Dora and Radha Raman Mishra

4D printing is based on 3D printing; however, it requires stimulus and stimulus-responsive materials. In the recent years, it has emerged as a potential technique that allows self-assembly, multi-functionality, and self-repair. These properties are achieved in 4D parts through the interaction of a stimulus with smart materials with subsequent evolution of 4D printed structures as a function of time. 4D printed structures are time-dependent, printer-independent, predictable, and intelligent enough to change their shape, properties, or functionality, in contrast to 3D printing. This chapter presents a comprehensive overview of 4D printing, including the process fundamentals, materials, techniques, challenges, and applications.

13.1 Introduction

The history of four-dimensional (4D) printing started with realization of three-dimensional (3D) printing. 3D printing refers to the method of producing a 3D geometry by depositing materials layer-by-layer using computer-aided design until the desired geometry is generated [1]. The idea of 3D printing was first suggested in 1983, when Chuck Hull patented the stereolithographic process [2]. Later, 3D printing or additive manufacturing (AM) processes such as fused deposition modelling, binder jetting, ink jet printing, direct ink writing, powder bed fusion, etc, were developed. The field emerged with major interest among researchers globally with possible applications in the automotive, biomedical, aeronautics, defence, construction, and space fields [3]. The rapid growth of AM processes is largely due to their potential to print products with a high degree of complexity with high precision, high speed, and at an economical cost. Prior to the year 2012, 3D printing was limited to the fabrication of static structures and was unable to serve the need for dynamic applications. In 2012 an extra dimension, i.e. time, was incorporated into the static 3D printed structures to demonstrate shape transformations over time by Professor Skyler Tibbits at MIT, USA. This marked the

doi:10.1088/978-0-7503-5351-9ch13

beginning of a new era in the field of 3D printing with the additional dimension of time, together referred to as 4D printing.

Several milestones have been achieved since the advent of 3D printing that led to significant developments in 4D printing. Currently, the 3D printing technology includes several forms of printing, such as stereolithography (SLA), selective laser sintering (SLS), direct ink writing (DIW), fused deposition modelling (FDM), jet 3D printing (3DP), electron beam melting (EBM), selective laser melting (SLM), powder bed fusion (PBF), material extrusion, binder jetting (BJ), material jetting (MJ), sheet lamination process, directed energy deposition (DED), and vat photo-polymerization processes [4]. There has been a multi-fold advancement in the research and development of 4D printing technology providing the ability to print complex structures. Currently, the major research effort in 4D printing is focused on the shape-changing capability, such as elongation, bending, and twisting of 4D printed materials. The progress of 4D printing can be compared with other emerging manufacturing technologies, such as smart dust, human augmentation, quantum computing, smart robots, and deep learning using the Gartner hype cycle (variations in expectations with time). It predicts the possibilities of these innovative technologies for practical applications with a highly competitive advantage. It has been predicted that 4D printing is, at present, in the innovation triggering stage and the expectations of 4D printing technique will increase with time [5]. Recently, 4D printing of various materials including metals, ceramics, and polymers has been reported through a multistage transformation [5]. In this chapter, the fundamentals of 4D printing, such as the process, materials, stimuli, and applications, are discussed.

13.2 Fundamentals of 4D printing

4D printing integrates time as the fourth dimension in 3D printing, enabling printed items to alter their shape or function in reaction to external stimuli such as temperature or light. The concept of time is important in 4D printing because it allows for the printing of stimulus-responsive materials that can respond to different stimuli at different times, resulting in the creation of complex structures and devices. The time-dependent shape-morphing behaviour of multi-material 4D printed structures is controlled by two kinds of time constants, which may be equal, large, or vanish depending on the stimulus and material used for printing, according to the third law of 4D printing [4]. A hypersphere is analogous to a fourth-dimensional sphere. Generally, a sphere appears three-dimensional but has a two-dimensional surface. In similar terms, a hypersphere has a three-dimensional surface and curves into four-dimensional space. The concept of a 4D hypersphere is not directly related to 4D printing technology. However, a number of sources suggest that the mathematical modelling and sequential stimulation employed in 4D printing can produce objects that can be transformed into a variety of forms, analogous to the transformation of hypersphere shapes [5]. In the future, hyperspheres can be utilized in simulations and the identification of shape-changing structures fabricated via 4D printing technique. The following sections provide the fundamentals associated with 4D printing, such as the process, stimuli, materials, and shape change programming.

13.2.1 Process

The 4D printing process follow three steps (pre-processing, fabrication, and post-processing) that are similar to 3D printing, whereas an additional step for programming of the printed part is used for achieving the required structure. Figure 13.1 illustrates schematically the processing steps that are involved in the fabrication of 4D printed objects. The pre-processing step includes material selection, design of the 3D printed objects, and part programming. Material selection is a critical step in the case of 4D printing. The materials used in 4D printing need to be very responsive to external stimuli and have some sort of shape memory effect. The selection of smart materials depends upon the type of AM machine which will be used to build the part as smart materials are available in different forms such as solid, liquid, and powder. Therefore, the design of the final 4D printed part depends on the properties of smart materials. The part design requires CAD modelling with subsequent conversion to extensions such as .STL, . IGES, etc. The use of an STL file with slicing software is more popular as an AM machine builds parts in a layer-by-layer manner by slicing of the components. Further, a G-code file containing all the data is assembled and sent to the AM machines. The build stage is carried on the stage of the AM machine. For smooth and accurate printing of the final part, parameters such as the printing speed, nozzle temperature, bed temperature, etc, need to be optimized before starting the actual printing. Some of the parts can be used directly after 3D printing; however, the use of support structures requires post-processing as it needs to be removed before the utilization of the printed part. To obtain the dynamic response and shape memory effect in the printed part, the part is actuated with some stimulus such as heat, temperature, and light (figure 13.1). The mechanisms involved in 4D printing through the interaction of the stimuli and smart materials are also indicated.

13.2.2 Stimuli

Stimuli play an important role in 4D printing to provide the fourth dimension for the 3D printed structure. With the application of one or more stimuli, the shape and function of the 3D built-in smart static structure can be changed. Stimuli are categorized into two categories, i.e. external and internal stimuli, based on their applications. Water/humidity, light, temperature, electric field, and magnetic field

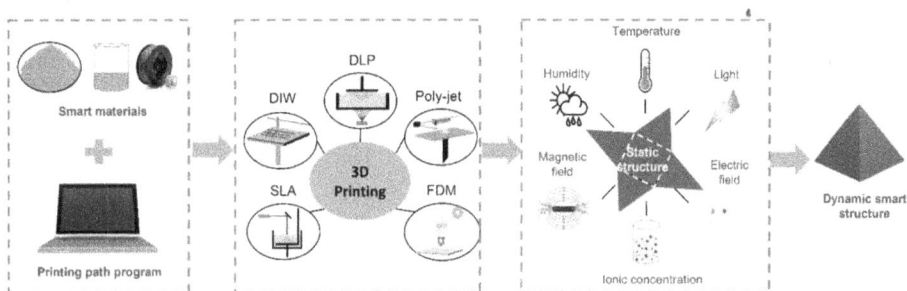

Figure 13.1. Schematic illustration of the steps involved in 4D printing process.

fall under the category of external stimuli, whereas cell traction force act as an internal stimulus. Stimuli can also be categorized based on the triggering phenomena —physical, chemical, biological, magnetic, electrical, etc, and/or a combination of any of these. Various details of the theory, mechanism, advantages, and disadvantages of all these stimuli are discussed below.

13.2.2.1 Water/humidity

Water/humidity is one of the first used stimuli in 4D printing. As water is omnipresent, the materials sensitive to water or humidity are of great interest and could find a wide range of applications in 4D printing. The static structure can be deformed inside water; subsequently, it regains its initial shape after drying. However, the quantity of expansion/compression must be properly regulated throughout the transition by ensuring the integrity of the printed components [6]. Hydrogels are hydrophilic materials and react extremely well with water to achieve 4D printing through a swelling and deswelling mechanism. The use of water or moisture as a stimulus is sanitary and practical, however, a delayed reaction time, poor mechanical strength after expansion, and deterioration of the structure after continuous swelling and deswelling limits their applications.

13.2.2.2 Light

Light has recently gained wide acceptance as a stimulus owing to its several characteristics such as instantaneous switching, precise focusing, and sustainable properties. Furthermore, the smart stable structure and undergo various transitions such as size or shape change, charge distribution, and photodimerization with light irradiation [7]. When exposed to light the smart material absorbs light and produces heat. The material starts to deform as a consequence of the heat, resulting in a change in the shape of the exposed component. Light, unlike heat and humidity, is an indirect stimulus since it does not induce the change directly [6]. It is mostly used for biomedicine application such as *in vivo* drug delivery.

13.2.2.3 Temperature

Temperature is a commonly used external stimulus for developing a shape deforming part. The shape change effect (SCE) and the shape memory effect (SME) are two mechanisms that are often responsible for the shape change of smart materials when temperature is used as a stimulus. SME-exhibiting materials are referred to as shape memory materials (SMMs), and these materials may be further divided into subgroups such as shape memory polymers (SMPs), shape memory alloys (SMAs), shape memory ceramics (SMCs), shape memory hybrids (SMHs), and shape memory gels (SMGs) [4]. The SMPs are used widely because of their easy printability. The SMPs are heated beyond their glass transition temperature (T_g) to achieve a metastable temporal shape upon cooling. The original shape can be achieved further if the temperature of the SMPs are raised up to their T_g [4, 6, 7]. However, as the temperature of the SMPs is raised above T_g, the SMPs absorb enough energy to become rubbery, and thus become prone to deformation.

13.2.2.4 Electric field/electricity

An electric field acts as an indirect stimulus and causes heating inside the materials which are responsive to the electric field due to resistive loss. Thus, electro-responsive materials undergo deformation once exposed to an electric field [4, 6]. The electric field is used in the biomedical field with some protective measures, otherwise it may cause cell death, membrane rupture, and local heating.

13.2.2.5 Magnetic field

Magneto-responsive materials exhibit deformation under magnetic fields. Magnetic nanoparticles can be doped in metals and polymers for achieving 4D printing of components. However, it works efficiently for lightweight metal and polymer parts [4]. The magnetic field's frequency in the biomedical field should also be within the medically safe range (50–100 kHz) to prevent tissue damage caused by temperature increases.

13.2.3 Materials

The selection of the printable material plays an important role in 4D printing. An appropriate material offers excellent functionality of the structure under an applied stimulus in terms of quick response time and rapid shape change ability. In recent years, AM processes have witnessed significant developments; however, quest for materials that have excellent printing properties for a targeted application is still on. It has opened a new area for research on printable smart materials. Some of the most widely used smart materials, their properties and application areas are shown in figure 13.2.

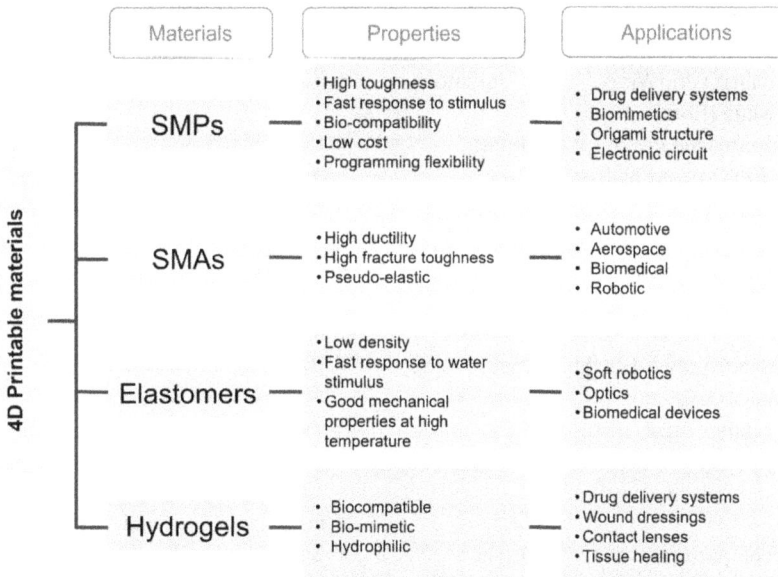

Figure 13.2. Types of 4D printing materials, their properties, and applications.

13.2.3.1 Shape-memory polymer (SMP)

In 1984 a polynorbornene based SMP was developed. Later, the Kuraray Company and the Asahi Company developed a higher glass transition temperature based SMP [8]. SMPs are the most preferred materials in 3D and 4D printing due to their high toughness, fast response to stimulus, low cost, programming flexibility, biocompatibility, and biodegradability. Additionally, SMPs are suitable for the aeronautical sector due to their high elastic modulus, ability to sustain higher strain, and better recovery rate. In SMPs the shape memory effect (SME) was actuated on its T_g, where the polymer changes from the glassy state to the elastic state, which facilitates the shape change [9]. SMPs offer a high response rate to stimulus such as heat, chemicals, and light. PLA is the most commonly used biodegradable SMP due to its availability and ease to form into shape using 3D printing machines such as FDM, SLA, and DLP. Current research in the area of SMPs is focused on the development of SMP-based composites. Some of the relevant research reported on SMPs is summarized in table 13.1.

Table 13.1. Commonly used SMPs in research.

S. No.	Material	Stimuli	Properties	References
1.	Rigid plastics (hydrophilic UV curable polymer)	Water (moisture)	• 150% increase in volume.	[10]
2.	Rigid plastics and a hydrophilic expandable polymer	Water (moisture)	• 200% increase in volume. • Elastic modulus: 40 MPa (dry) and 5 MPa (wet)	[11]
3.	Tangoblack (elastomeric matrix) and Vero white (fibre)	Light—photo initiators	• T_g of matrix: −5 °C • T_g of fibre: −47 °C	[12]
4.	Polyethylene glycol (PEG)-based hydrogel bilayers	Light	• PEG bilayers are triggered by the presence of different swelling ratios (5.06; 9.67; 35.09).	[13]
5.	Aqueous droplets in oil	—	• Connection through lipid bilayers, creating a cohesive material.	[14]
6.	Alginate/poly NIPAAm ionic covalent entanglement (ICE) gel	Heat	• Poly NIPAAm causes reversible volume changes and toughens the matrix.	[15]
7.	Thermoplastic elastomer	—	• Viscoelastic behaviour helps in shape change and SME.	[16]
8.	A composite hydrogel ink	Light	• Shows high stiffness. • No reversibility was noticed.	[17]

			• Reversibility in hot and cold water can be achieved by replacing poly(N,N-dime-thylacrylamide) matrix with stimuli-responsive poly (N-isopropylacrylamide).	
9.	Polylactic acid (PLA) and paper	Heat	• T_g: –60 °C. • E (elastic modulus): 3.5 GPa.	[18]
10.	Hygromorphic biocomposite	Moisture	• Hydro-elastic behaviour observed with high porosity microstructures (–20% volume percent).	[19]
11.	Benzyl methacrylate (BMA)	Light	• This network of polymers is beneficial for a projection micro stereolithography based AM system.	[20]
12.	Poly (2-vinylpyridine) and 12% wt of ABS	pH responsive (chemical) stimulus	• Mechanical properties were improved.	[21]
13.	Semi-crystalline methacrylated polycaprolactone (PCL)	Heat	• Shows enhanced rheological and viscoelastic properties. • Vitamin E hinders cross-linking of polymers.	[22]

13.2.3.2 Shape-memory alloys (SMAs)

SMAs are smart metal alloys as they transform into multiple phases once exposed to stimuli such as temperature, stress, or a combination of both. At low temperatures, SMAs exist in the martensitic phase which is ductile and easily deformable. As the temperature increases, SMAs achieve the austenitic phase that reduces the stress intensity factor to achieve higher fracture toughness [8]. Pseudo-plasticity and the shape memory effect are two characteristics that are found in SMAs. The SME in the SMAs is governed by the reversible transition between the austenitic and martensitic phases. The most commonly used SMAs for 4D printing are nitinol (a nickel-titanium alloy consisting of 50% of nickel and 50% titanium), Ni-Mn-Ga and Cu-Al-Ni [23]. SMAs have found applications in the automotive, aerospace, biomedical, and robotic sectors due to their high modulus and high flexibility. The drawbacks of using SMAs in 4D printing are complex programming, high density, non-biocompatibility, and high material costs. The most used SMAs and relevant research reported are presented in table 13.2.

13.2.3.3 Hydrogels

Hydrogels are composed of multiple layers of hydrophilic monomers. Upon exposure to stimuli, the volume of the printed hydrogel structure increases drastically. The hydrophilic nature of hydrogels offers significant expansion upon

Table 13.2. Commonly used SMAs in research.

S. No.	Material	Stimuli	Properties	References
1.	SMA wires	Mechanical force	• Debonding strength: NiTiFe (54.64 GPa) > NiTiCu (36.62 GPa) > NiTi (11.16 GPa)	[24]
2.	SMA	Heat	• Total energy consumed for actuation is 37.8 kJ	[25]
3.	TiNi	Heat	• Ultimate tensile strength: 776 MPa • Elongation: 7.2 • Shape recovery ratio: 98.7%	[26]
4.	Ni-Mn-Ga	Heat	• A rotational magnetic field of 0.97 T was applied. • Highest porosity: 70,43	[27]
5.	SMA	Electric current and heat	• After actuation, the height of a single hollow pocket increased by 325% and a decrease of 14% in the width was reported.	[28]

water absorption. Therefore, water can be used as a stimulus for 4D printing of hydrogels. Hydrogels have been found to be suitable for applications such as drug delivery systems, wound dressings, contact lenses, and tissue healing due to their biocompatibility and biomimetic nature. However, the low mechanical strength and low Young's modulus of the printed components results in brittle and fragile components and limits their uses for 4D printing. Some commonly used hydrogels and relevant research are presented in table 13.3.

13.2.4 Shape change programming

In a smart material, a steady release of stored strain energy causes a shape or a functional change during the printing process. Once the stimulus is retracted, the recognition of the transformation depends on the material's response to it. Once the triggering stimulus is terminated, the smart material returns to its principal shape. Shape (permanent/temporary) programming of a build structure can be achieved by various techniques. The most common technique is inducing strain in a 3D printed part by heating it beyond the transition temperature. This programming offers easy prediction and repetition due to the possibility to measure the force and induced strain; however, the 4D printed structure becomes a permanent structure. On the other hand, programming of SMPs can be done in the pre- and post-printing stage. The pre-strain induced 4D printing process provides accurate measurement of strains and can be used in precise shape change programming in the structure's temporary state. It is known as non-manual programming. Such techniques that dictate the shape change mechanism in the activated SME are called shape programming of 4D printed structures. The programming can be categorized by the nature of the shape programmed into the structure. Also, different shape

Table 13.3. Commonly used hydrogels in research.

S. No.	Materials	Stimuli	Properties	References
1.	Poly(NIPAM-co-DMAPMA)/clay hydrogel	Moisture and heat	• Bi-layered hydrogel: water bath • Deformation temperature: 80 °C • Swelling rate: top layer 52.09; bottom layer 1.11 ± 0.08	[29]
2.	Alginate (Alg) and methylcellulose (MC)	pH and moisture-based stimulus	• Deionized water and 0.1 M $CaCl_2$ solution was utilized. • A 3D bioprinter was utilized to construct diverse dynamic architectures.	[30]
3.	Alginate (ALG), methylcellulose (MC), and polyacrylic acid (PAA)	Magnetic field	• Ferrofluid obtained with a pH −7. • Experimental temperature: 80 °C	[31]
4.	Agarose/PAAm: agarose + AAm + crosslinker + photoinitiator + laponite nanoclay	Heat- and light-based stimuli	• Gel softened above T_g: −95 °C • Increase in temperature reduces the storage modulus (E') • Ratio of elastic modulus and dissipated energy: over 80%	[32]
5.	Alginate and hyaluronic acid	pH based actuation	• Cell viability in self-folded tubes formed (films 3DP at 80/40 mm min^{-1} on glass surfaces): 97%/96% (1 day), 98%/96% (2 days), and 98%/97% (7 days) • Resulting tubes: support cell survival (at least 7 days without any decrease in cell viability).	[33]

Table 13.4. Shape programming of SMP-printed materials.

S. No.	Programming	Stimuli	Effect
1.	Bending	Strain-inducing bending	Recovery from permanent shape
2.	Twisting	Strain-inducing straightening	Recovery to printed spring structure
3.	Unfolding	Edge folding	Recovery to shape
4.	Sequential shape change	Varying section thickness	Activation of sequential hinges
5.	Elevation	Activation	Elevation of auxetic patterns
6.	Local bending	Light-induced heating	Activation of certain sections of the structure
7.	Concave	Varying print temperature to control induce printing strain	Concentric printing pattern of SMP
8.	Expand to fit a shape	—	Hexagonal patterns recover full expansion shape
9.	Helix	Contract strain	Recovery to printed helix pattern
10.	Non-uniform	Control through grayscale values	Cross-linking of polymer

programming is needed while using a single or multiple material. Table 13.4 presents the shape programming of SMP-printed materials.

13.3 Applications

The 4D printing technology offers dynamically smart parts, greater prototyping freedom, and employs smart materials. 4D printing is an emerging area, however, it has attracted significant interest from various industries including biomedical, automotive, aerospace, defence, and electronics (figure 13.2). The need for organ transplants has grown considerably over the past few years. Bio-printing of organs is used in tissue engineering to produce high-accuracy cell placement, high-density tissue, and huge tissue designed products. 4D bio-printing allows the fabrication of biological constructs with shape-shifting properties in response to an external stimulus. In addition, with the development various stimuli responsive hydrogels, 4D printing has significant scope in drug delivery system. The current industrial technical trend is toward more human-friendly devices that exhibit flexibility and mobility. Therefore, the demand for stretchable and elastic devices and smart surfaces like human skin has been on the rise. 3D and 4D printing have the capability to realize such devices that have the ability to allow integration of electronics with our routine lives but could also be indispensable in medical diagnoses such as health monitoring, the utilization of artificial skins, and implantable bioelectronics. The application of a 4D printed self-changing polymer as a soft electronic actuating device was reported recently. This device has reduced design complexities and costs, which serves the purpose of a smart material application [34]. Morphing structures, widely applicable in the aerospace industry, have the capacity to change their geometric properties in response to temperature or moisture. The use of 4D printed materials has been reported in defence and protection [34]. Smart automotive coatings for improved corrosion protection are being developed to detect and to respond to a humid environment by changing their build structure. Another application of 4D printing is the smart soldier's uniform which can alter its structure, camouflage itself, and efficiently protect a human body from deadly gas or shrapnel upon contact.

13.4 Challenges

In spite of the huge popularity of 4D printing due to the possibility of printing complex dynamic structures, there exists some challenges which hinder faster industrial adoption. The main challenge is associated with the limited availability of smart materials and the need for the development of new smart materials. For example, an issue associated with materials is their form, as the material in the FDM process should be a solid wire, whereas for polyjet printing the material should be in the liquid state. Moreover, thermally responsive smart material cannot be build using the FDM process as this process needs a high processing temperature for melting and indirectly affects the material properties [4, 35]. Another, challenge is with the limited availability of AM processes that are suitable for 4D printing. Further, new materials and processes need to be researched and established. Specific material printing is only possible on a suitable printer as different printers require the material to be in different conditions [35]. Moreover, the activation method, part strength, shape change speed, etc, of 4D printed objects are strongly affected by the properties of the stimulus-responsive materials [36]. Some materials tends to degrade due to continuous deformation while materials such as hydrogels degrade through repetitive wetting and drying [4]. 4D printed smart structures [4] depend upon material properties and the printing process. Understanding the interactions of processes in multi-materials before and after printing can aid in optimizing the performance of printed devices [35].

13.5 Future perspective

As a relatively new technique, the majority of research on 4D printing is still in the R&D phase. However, more research and development can be imposed to make 4D printing more economical and user-friendly in the future, which might result in a broader range of applications of 4D printed products [37].

This will require interdisciplinary research and development for 4D printing as well as technological advancements in a variety of fields, such as new design and modelling tools, enhanced hardware for 3D printing, and material science for smart materials. In particular, multi-material 3D printing technologies and high-speed, high-resolution printers will compete to meet the demands of mass manufacturing of materials and devices with multi-scale, complicated geometries [36]. 4D printing's ability to scale up or down means it can be used to create items with complex geometries that would be impossible to print with traditional 3D printers. New materials and stimuli might need to be explored further to enhance the usefulness of 4D printed parts [38].

From a material perspective, in order to ensure 4D printing's success in the future, the development of a new and customizable material which can respond to a variety of stimuli needs to be explored. In addition to this, for different 4D printing techniques and materials, a 4D printing software needs to be developed, which will consider all the parameters and mechanism such as the shape-changing mechanism, the geometrical and structural requirements of the product, etc [8].

References

[1] Mercado F and Arciniegas A J R 2020 Additive manufacturing methods: techniques, materials, and closed-loop control applications *Int. J. Adv. Manuf. Technol.* **109** 17–31

[2] Hull C W 1986 Apparatus for production of three-dimensional objects by stereolithography *US Patent Specification* 4575330

[3] Mostafaei A, Elliott A M, Barnes J E, Li F, Tan W, Cramer C L, Nandwana P and Chmielus M 2021 Binder jet 3D printing—process parameters, materials, properties, modeling, and challenges *Prog. Mater Sci.* **119** 100707

[4] Ahmed A, Arya S, Gupta V, Furukawa H and Khosla A 2021 4D printing: fundamentals, materials, applications and challenges *Polymer* **228** 123926

[5] Joharji L, Mishra R B, Alam F, Tytov S, Al-Modaf F and El-Atab N 2022 4D printing: a detailed review of materials, techniques, and applications *Microelectron. Eng.* **265** 111874

[6] Chu H, Yang W, Sun L, Cai S, Yang R, Liang W, Yu H and Liu L 2020 4D printing: a review on recent progresses *Micromachines* **11** 796

[7] Lui Y S, Sow W T, Tan L P, Wu Y, Lai Y and Li H 2019 4D printing and stimuli-responsive materials in biomedical aspects *Acta Biomater.* **92** 19–36

[8] Joshi S, Rawat K, Rajamohan K C V, Mathew A T, Koziol K, Kumar Thakur V and Balan A S S 2020 4D printing of materials for the future: opportunities and challenges *Appl. Mater. Today.* **18** 100490

[9] Alshebly Y S, Nafea M, Mohamed Ali M S and Almurib H A F 2021 Review on recent advances in 4D printing of shape memory polymers *Eur. Polym. J.* **159** 110708

[10] Tibbits S, McKnelly C, Olguin C, Dikovsky D and Hirsch S 2013 4D printing and universal transformation *Proceedings of the 34th Annual Conference of the Association for Computer Aided Design in Architecture* pp 539–48

[11] Raviv D *et al* 2014 Active printed materials for complex self-evolving deformations *Sci. Rep.* **9** 7422

[12] Ge Q, Dunn C K, Qi H J and Dunn M L 2014 Active origami by 4D printing *Smart Mater. Struct.* **23** 094007

[13] Jamal M, Kadam S S, Xiao R, Jivan F, Onn T, Fernandes R, Nguyen T D and Gracias D H 2013 Bio-origami hydrogel scaffolds composed of photocrosslinked PEG bilayers *Adv. Healthc. Mater.* **2** 1142–50

[14] Villar G, Graham A D and Bayley H 2013 A tissue-like printed material *Reports* **340** 48–53

[15] Bakarich S E, Iii R G and Spinks G M 2015 4D Printing with mechanically robust, thermally actuating hydrogels *Macromol. Rapid Commun.* **36** 1211–7

[16] Mutlu R, Alici G and Spinks G 2015 Effect of flexure hinge type on a 3D printed fully compliant prosthetic finger in *IEEE Int. Conf. Adv. Intell. Mechatronics* (Piscataway, NJ: IEEE) pp 1–6

[17] Gladman A S, Matsumoto E A, Nuzzo R G, Mahadevan L and Lewis J A 2016 Biomimetic 4D printing *Nat. Mater.* **15** 413–8

[18] Zhang Q, Zhang K and Hu G 2016 Smart three-dimensional lightweight structure triggered from a thin composite sheet via 3D printing technique *Nat. Publ. Gr.* **8** 22431

[19] Le Duigou A, Castro M, Bevan R and Martin N 2016 3D printing of wood fibre biocomposites: from mechanical to actuation functionality *Mater. Des.* **96** 106–14

[20] Ge Q, Sakhaei A H, Lee H, Dunn C K, Fang N X and Dunn M L 2016 Multimaterial 4D printing with tailorable shape memory polymers *Sci. Rep.* **11** 31110

[21] Nadgorny M, Xiao Z, Chen C and Connal L A 2016 Three-dimensional printing of pH-responsive and functional polymers on an affordable desktop printer *ACS Appl. Mater. Interfaces* **8** 28946–54

[22] Zarek M, Mansour N, Shapira S and Cohn D 2017 4D printing of shape memory-based personalized endoluminal medical devices *Macromol. Rapid Commun.* **38** 1600628

[23] Pingale P, Dawre S, Dhapte-Pawar V, Dhas N and Rajput A 2022 Advances in 4D printing: from stimulation to simulation *Drug Deliv. Transl. Res.* **13** 164–88

[24] Sreesha R B, Ladakhan S H, Mudakavi D and Adinarayanappa S M 2022 An experimental investigation on performance of NiTi-based shape memory alloy 4D printed actuators for bending application *Int. J. Adv. Manuf. Technol.* **122** 4421–36

[25] Testoni O *et al* 2021 A 4D printed active compliant hinge for potential space applications using shape memory alloys and polymers *Smart Mater. Struct.* **30** 085004

[26] Lu H Z, Yang C, Luo X, Ma H W, Song B, Li Y Y and Zhang L C 2019 Ultrahigh-performance TiNi shape memory alloy by 4D printing *Mater. Sci. Eng.* A **763** 138166

[27] Caputo M P, Berkowitz A E, Armstrong A, Müllner P and Solomon C V 2018 4D printing of net shape parts made from Ni-Mn-Ga magnetic shape-memory alloys *Addit. Manuf.* **21** 579–88

[28] Akbari S, Hosein A, Panjwani S, Kowsari K, Serjouei A and Ge Q 2019 Sensors and actuators a: physical multimaterial 3D printed soft actuators powered by shape memory alloy wires *Sens. Actuators* A **290** 177–89

[29] Li Y, Zheng W, Li B, Dong J, Gao G and Jiang Z 2022 Physicochemical and engineering aspects double-layer temperature-sensitive hydrogel fabricated by 4D printing with fast shape deformation *Colloids Surf.* A **648** 129307

[30] Lai J, Ye X, Liu J, Wang C, Li J, Wang X, Ma M and Wang M 2021 4D printing of highly printable and shape morphing hydrogels composed of alginate and methylcellulose *Mater. Des.* **205** 109699

[31] Nizio M, Szymczyk-zi P, Simi J, Junka A, Shavandi A and Podstawczyk D 2022 4D printing of patterned multimaterial magnetic hydrogel actuators *Addit. Manuf.* **49** 1–14

[32] Guo J, Zhang R, Zhang L and Cao X 2018 4D printing of robust hydrogels consisted of agarose nano fibers and polyacrylamide *ACS Macro Lett.* **7** 442–6

[33] Kirillova A, Maxson R, Stoychev G, Gomillion C T and Ionov L 2017 4D biofabrication using shape-morphing hydrogels *Adv. Mater.* **29** 1–8

[34] Wu J J, Huang L M, Zhao Q and Xie T 2018 4D printing: history and recent progress *Chinese J. Polym. Sci.* **36** 563–75

[35] Fu P *et al* 2022 4D printing of polymers: techniques, materials, and prospects *Prog. Polym. Sci.* **126** 101506

[36] Kuang X, Roach D J, Wu J, Hamel C M, Ding Z, Wang T, Dunn M L and Qi H J 2019 Advances in 4D printing: materials and applications *Adv. Funct. Mater.* **29** 1–23

[37] Haleem A, Javaid M, Singh R P and Suman R 2021 Significant roles of 4D printing using smart materials in the field of manufacturing *Adv. Ind. Eng. Polym. Res.* **4** 301–11

[38] Raina A, Haq M I U, Javaid M, Rab S and Haleem A 2021 4D printing for automotive industry applications *J. Inst. Eng. D* **102** 521–9

3D Printed Smart Sensors and Energy Harvesting Devices
Concepts, fabrication and applications
Sanket Goel and Sohan Dudala

Chapter 14

3D printing: challenges and future opportunities

Sohan Dudala and Sanket Goel

This chapter briefly provides an overview of essential considerations in 3D printing workflow, encompassing design tools, fabrication processes, material innovations, and post-processing techniques. The chapter begins with computer-aided design (CAD) tools for creating 3D models and highlights the role of slicing software in bridging digital designs with printed products. Challenges in fabrication processes are explored alongside advancements in materials like conductive and graphene-based filaments. The chapter highlights post-processing methods and future opportunities to enhance scalability, sustainability, and performance in 3D printing.

The development of 3D printing technology involves a broad spectrum of tools, techniques, and processes. Each step is vital to achieving high-quality products, from the initial design phase using CAD software to the final stages of post-processing. This chapter briefly discusses a few aspects of 3D Printing—design tools, fabrication processes, material innovations, and post-processing techniques—while highlighting current challenges and future opportunities.

14.1 Design tools for 3D printing

The first step in the entire 3D printing process value chain is the design of 3D models, which is achieved using computer-aided design (CAD) tools. The CAD files, generated using CAD tools, comprise the set of instructions or regulations dictating the operations of 3D printers. The process involves the determination of the required quantity of material to be deposited and the appropriate location for its placement. The CAD files are subsequently transmitted in more appropriate files (like .stl or g-codes) to the 3D printer for further processing. The tools employed for these processes have been extensively discussed in the chapter by Arya *et al*.

The utilization of computer-aided design (CAD) tools for design and modelling is an inherent and essential aspect of 3D printing technology. A multitude of feature-rich software are available to meet the design requirements for 3D printing purposes. A few noteworthy options are 3ds Max (Autodesk), AC3D (Invis Limited), Solid

doi:10.1088/978-0-7503-5351-9ch14
14-1

Works (Dassault Systèmes SE), Creo (PTC), amongst others. Although highly versatile and capable of designing intricate features, these software have limitations in terms of accessibility due to high procurement/subscription costs limiting accessibility to universities, academic or commercial research institutions and corporates. With the expanding accessibility of 3D printing hardware, new 3D design tools, such as LibreCAD, Tinker CAD, Fusion 360 (Autodesk), FreeCAD, SketchUp, LeoCAD, etc, have also emerged which are either open source or free for limited use. Notably, a few of these free or open-source software have emerged as online cloud-based CAD tools that enable use with basic computing hardware. Even though a multitude of options are now available for CAD designing, an essential shortcoming of their use in 3D printing is the lack of integrated workflow. These tools primarily focus on effective designing without often considering the limitations of the 3D printing process technology. A design tool has the potential to generate a very advanced and complex CAD model. Still, it may not be compatible with the technology and process involved in 3D printing. The responsibility lies with the user to thoroughly analyse all relevant factors and make a cautious decision about design considerations, perhaps resulting in underutilizing the 3D printing capabilities of the available hardware. One possible approach to address this issue involves the incorporation of 3D printability assessment directly into CAD software. The integration should encompass a comprehensive evaluation of 3D printability, taking into account the specific make and model of the printer. A shared database of commercially available printers should support this evaluation method.

The slicing software or the slicers link CAD designs with 3D printed products. Slicing software is an intermediary between the digital model (generated by CAD tool) and fabricated product by the 3D printer. It converts the digital model (often in .stl .3df, .obj, .amf or other related formats) into instructions for the printer to understand, such as G-code. The slicing software commonly manages several aspects of the 3D printing process, including material parameters, layer height, and support structure requirements. The slicer performs the crucial task of dividing the 3D object into two-dimensional layers for printing, utilizing all relevant process parameters. The slicing software is commonly bundled with the printer package to ensure optimal performance. However, there are alternative solutions available that offer cross-compatibility, such as Cura, Creality, ideaMaker, Slic3r, and others. The accessibility of this slicing software is no longer an issue, as was the case with proprietary designing tools. However, the main challenge with slicing software is the non-uniformity of the final product with different slicing software. A significant source that can be utilized to exemplify this point is the research conducted by Šljivic et al [1]. The surface quality, dimensional accuracy, production time, and raw material consumption were compared for an identical CAD design using the same hardware across three slicing software programs: Simplify3D, Cura, and Slic3r 3D. Simplify3D was the preferred recommendation of Šljivic et al [1] followed by Cura and Slic3r 3D. This observation supports the argument that there are notable variations in dimensional and surface accuracy across multiple slicers, posing a challenge that must be addressed to enhance the repeatability and reliability of 3D printing.

14.2 3D printing process technology

While discussing 3D printing, fabrication refers to the layer-by-layer deposition or curing of material depending on the type of 3D printing process technology. Various types of 3D printing techniques have been discussed in the chapter dealing with types of 3D printing techniques. While the type of 3D printing process technology is essential, the materials used for the process and the supports needed in case of complex geometrical design are equally important. Challenges exist in all of these subsets of fabrication in 3D printing, which need to be discussed to find an amicable solution.

The most popular 3D printing process is material extrusion-based—fused deposition modelling (FDM) or fused filament fabrication (FFF). In terms of popularity and consumer acceptability, FDM is closely followed by stereolithography (SLA), polymer resin-based 3D printing. FDM is a popular 3D printing technology but faces several challenges. Key issues include poor surface quality, insufficient mechanical properties of printed parts and low dimensional accuracy [2]. Warpage and poor adhesion have also been reported with specific categories of material [3]. While this is the case with most 3D printing technologies, the effect of process parameters like layer thickness, build orientation, and print speed have a higher influence on part quality with FDM [2]. The future opportunities lie in improving printing equipment to enhance 3D printed part characteristics. Advancements have been made to improve the productivity of FDM 3D printers with technologies like independent dual (or multi) extruder printers. Independent multi-extruder printers have significantly improved productivity and offered features like mirror mode printing, which are a step towards enhancing the scalability of 3D printing for manufacturing. Key hardware improvements are needed to address the limitations of resolution and repeatability of the printing process using FDM. Most FDM 3D printers rely on a few types of NEMA servo motors. These motors have been pivotal in making 3D printers affordable and offer good performance at the price point. However, the time has come to look beyond them. Alternatives exist, such as brushless DC motors with encoders, closed-loop stepper motors, and piezoelectric motors. Currently, affordability, when compared with NEMA steppers, limits the use of these alternatives, but as applications of FDM 3D printing evolve for highly specific use cases, they will be the options equipment manufacturers will look to.

The 3D printing technology that has picked up significant popularity in the recent past is SLA. Apart from being a popular consumer printer for prototyping owing to its higher resolution and superior surface finish, it is also used in specialized applications. Most challenges with SLA printing are associated with materials. Due to the UV and temperature-sensitive nature of the resins, extreme care is to be taken in storing and handling these resins. The shelf life of resins is also significantly shorter compared to filaments. Two common issues every user of an SLA printer can attest to are the strong (often unpleasant) odour of resins and the difficulty in cleaning resins in cases of accidental spills. Most SLA printers are also restrictive in terms of materials use. They are limited to those supplied by the manufacturer in proprietary cartridge/tank-based formats. The challenges with the SLA printing process include improving part performance, printing speed and reducing distortion and defect formation [4, 5]. These can be addressed by a combination of improvements—both in materials and hardware. Better laser systems with raster mode capabilities are one method by which printing speeds can be enhanced. The ongoing improvements in SLA 3D printing technology are promising, with applications ranging from fabricating complex structures like ceramic cores for turbine blades to biological applications such as tissue scaffolds for regenerative medicine [4, 6]. Opportunities in SLA printing also lie in developing innovative materials, such as conductive photopolymers and photopolymer resins with enhanced mechanical and functional properties.

14.3 Materials and applications

Materials are quintessential in the 3D printing process. They have been discussed in detail throughout this book. While quite a few options are now available for all types of 3D printing techniques, challenges are still associated. Conductive materials such as carbon-doped thermoplastics and copper meshes have been used with various 3D printing techniques, including FDM and poly-jetting, to develop sensors [7–9]. These sensors can detect mechanical flexing and capacitance changes. The integration of sensing capabilities into 3D printed structures has enabled the production of custom sensing devices and objects with embedded sensors in a single build process [9–12].

Graphene-based 3D printing filaments are another material category that has shown promising use in sensors and energy harvesters. Acquah *et al* [13] discussed in 2016 the commercial availability of carbon nanotube and graphene-infused 3D printing filaments, highlighting their potential for producing conductive composites. Graphene-based filaments have seen widespread use in applications needing improved mechanical properties [14], energy storage applications [15], fuel cells [16] and applications requiring thermal stability with electrical conductivity [17]. Graphene-based filaments have been reported in the past for applications of 3D printed sensors spanning across physical, chemical, and biological domains, with potential uses in biomedical sensing, human interface devices, and environmental monitoring [7, 18, 19]. Challenges remain with graphene-based filaments in achieving optimal dispersion and preventing nozzle clogging during printing [14]. These challenges are the primary reason impacting their commercial availability. Better material compositions and improved printheads or hot ends are needed to address these challenges and revitalise the use of graphene filaments.

Recent advancements have expanded the range of printable materials, including biobased materials, smart materials, ceramics, and composites. These developments offer new possibilities in the biomedical, food, and textile industries. Future opportunities lie in developing multifunctional materials to design complex systems inspired by Nature. As the field continues to evolve, adapting to changing materials and product requirements will be essential for realizing the full potential of 3D printing in manufacturing and product development. With the growth of the 3D printing market, there is an increasing focus on sustainable materials and processes to address the rising concern of plastic waste [20]. The challenge is more so with SLA resins. While most filaments are thermoplastics, resins are thermosetting, significantly reducing their recyclability. Some efforts, such as resin reclamation and downcycling, have been undertaken for SLA printed products, but they are just short-term solutions. To ensure long-term stability, extensive research and development are needed in SLA resin recycling.

14.4 Post processing

One key limitation often associated with 3D printing is the need for post-processing. While some simple jigs and fixtures for sensors and energy harvesters may not require post-processing, strength and surface-dependent applications need post-processing. The need for post-processing may also arise due to various other causes— appearance, functionality, ergonomics, use case or application. In cases where 3D-printed parts are used as end products, post-processing is often done to improve aesthetics and hide the layered remnants of the process. Arguably, one of the simplest forms of post-processing for aesthetic considerations is painting or coating. In addition to paints, advanced coatings are commercially available to improve aesthetics. One such product is Smooth-On XTC-3D, a two-part coating liquid for smoothing and finishing 3D printed parts [21]. It is primarily used for FDM printed parts. In addition to conventional brush-based coatings or paints, spray formulations are also used for a more uniform finish. Solvent exposure can also be used to improve surface properties. These solvents play an essential role when 3D printed products are used for microfabrication or as moulds in soft-lithography processes [22]. Post-processing may also be needed to improve the performance of a 3D-printed device. For instance, dimethylformamide (DMF) is often used to activate surfaces of graphene-PLA filament-based prints in applications requiring enhanced electro-active surface area [16].

While post-processing is often associated as an additional step with filament-based 3D printing or sintering-based 3D printing, it is essential in SLA-based 3D printing. Washing with a suitable solvent (isopropyl alcohol) and phot-thermo curing are essential steps in SLA-based 3D printing. While these processes are essential for decent surface and strength properties, their deployment is resource-intensive and often unpleasant. Improvements can be made to automate the printing, washing and curing processes for SLA-based 3D printing to offer a more pleasant user experience.

14.5 Future outlook

The future of 3D printing is incredibly promising. The focus in the previous decade was to make 3D printing affordable and accessible. The focus is now bound to change towards scalability, sustainability and enhanced capabilities. The anticipated developments in the 3D printing sphere are expected to significantly alter the development and prototyping of sensors and energy harvesters. Advances in materials are expected to lead this by employing functional materials (piezoelectric or thermoelectric materials) being integrated directly into 3D-printed structures. Multimaterial printing, specifically beyond filament-based printing techniques, is expected to be a significant development. Work is already underway to facilitate the direct integration of electronics, such as antennas, circuits, and batteries, into 3D-printed structures. Combining 3D printing with traditional manufacturing techniques will enable the creation of hybrid devices with optimal performance.

References

[1] Šljivic M, Pavlovic A, Kraišnik M and Ilić J 2019 Comparing the accuracy of 3D slicer software in printed enduse parts *IOP Conf. Ser.: Mater. Sci. Eng.* **659** 012082

[2] Dey A and Yodo N 2019 A systematic survey of FDM process parameter optimization and their influence on part characteristics *J. Manuf. Mater. Process* **3** 64

[3] Verma N, Awasthi P, Gupta A and Banerjee S S 2023 Fused deposition modeling of polyolefins: challenges and opportunities *Macromol. Mater. Eng.* **308** 2200421

[4] Li J, Boyer C and Zhang X 2022 3D printing based on photopolymerization and photo-catalysts: review and prospect *Macromol. Mater. Eng.* **307** 2200010

[5] Manapat J Z, Chen Q, Ye P and Advincula R C 2017 3D printing of polymer nano-composites via stereolithography *Macromol. Mater. Eng.* **302** 1600553

[6] Chartrain N A, Williams C B and Whittington A R 2018 A review on fabricating tissue scaffolds using vat photopolymerization *Acta Biomater.* **74** 90–111

[7] Shemelya C *et al* 2013 3D printed capacitive sensors *Proc. IEEE Sens.* **2013** 1–4

[8] Dijkshoorn A *et al* 2018 Embedded sensing: Integrating sensors in 3-D printed structures *J. Sens. Sens. Syst.* **7** 169–81

[9] Leigh S J, Bradley R J, Purssell C P, Billson D R and Hutchins D A 2012 A simple, low-cost conductive composite material for 3D printing of electronic sensors *PLoS One* **7** e49365

[10] Puneeth S B and Goel S 2019 Novel 3D printed microfluidic paper-based analytical device with integrated screen-printed electrodes for automated viscosity measurements *IEEE Trans. Electron Devices* **66** 3196–201

[11] Puneeth S B, Puranam S A and Goel S 2019 3D printed integrated and automated electro-microfluidic viscometer for biochemical applications *IEEE Trans. Instrum. Meas.* **68** 2648–55

[12] Puneeth S B, Hithesh H L and Goel S 2021 Electro-microfluidic viscometer with integrated microcontroller and pumping system for point-of-care biosensing applications *IEEE Instrum. Meas. Mag.* **24** 23–8

[13] Acquah S F A *et al* 2016 carbon nanotubes and graphene as additives in 3d printing *Carbon Nanotubes: Current Progress of their Polymer Composites* (Books on Demand)

[14] Aumnate C, Pongwisuthiruchte A, Pattananuwat P and Potiyaraj P 2018 Fabrication of ABS/graphene oxide composite filament for fused filament fabrication (FFF) 3D printing *Adv. Mater. Sci. Eng.* **2018** 2830437

[15] Foster C W *et al* 2017 3D printed graphene based energy storage devices *Sci Rep.* **7** 1–11

[16] Rewatkar P and Goel S 2019 Next-generation 3D printed microfluidic membraneless enzymatic biofuel cell: cost-effective and rapid approach *IEEE Trans. Electron Devices* **66** 3628–35

[17] Jayanth N, Senthil P and Mallikarjuna B 2022 Experimental investigation on the application of FDM 3D printed conductive ABS-CB composite in EMI shielding *Radiat. Phys. Chem.* **198** 110263

[18] Rao C H, Avinash K, Varaprasad B K S V L and Goel S 2022 A review on printed electronics with digital 3D printing: fabrication techniques, materials, challenges and future opportunities *J. Electron. Mater.* **51** 2747–65

[19] Pal A, Amreen K, Dubey S K and Goel S 2021 Highly sensitive and interference-free electrochemical nitrite detection in a 3D printed miniaturized device *IEEE Trans. Nanobiosci.* **20** 175–82

[20] Maines E M, Porwal M K, Ellison C J and Reineke T M 2021 Sustainable advances in SLA/DLP 3D printing materials and processes *Green Chem.* **23** 6863–97

[21] Haidiezul A H M, Aiman A F and Bakar B 2018 Surface finish effects using coating method on 3D printing (FDM) parts *IOP Conf. Ser.: Mater. Sci. Eng.* **318** 012065

[22] Dudala S, Dubey S K and Goel S 2019 Fully integrated, automated, and smartphone enabled point-of-source portable platform with microfluidic device for nitrite detection *IEEE Trans. Biomed. Circuits Syst.* **13** 1518–24

www.ingramcontent.com/pod-product-compliance
Lightning Source LLC
Chambersburg PA
CBHW080529220326
41599CB00032B/6257